Fadhil J. Kadhim
Mohamed Abdel-Raheem
Bayader Abd al-Jabbar Ali

Sustainable partnership between ornamental plants and honeybees

Fadhil J. Kadhim
Mohamed Abdel-Raheem
Bayader Abd al-Jabbar Ali

Sustainable partnership between ornamental plants and honeybees

Noor Publishing

Cover image: www.ingimage.com

Publisher:
Noor Publishing
is a trademark of
Dodo Books Indian Ocean Ltd. and OmniScriptum S.R.L publishing group

120 High Road, East Finchley, London, N2 9ED, United Kingdom
Str. Armeneasca 28/1, office 1, Chisinau MD-2012, Republic of Moldova, Europe
Managing Directors: Ieva Konstantinova, Victoria Ursu
info@omniscriptum.com

Printed at: see last page
ISBN: 978-620-2-43873-5

Sustainable partnership between ornamental plants and honeybees

By

Dr. Fadhil J. Kadhim

Prof. Dr. Mohamed Abdel-Raheem

Assist. Teacher. Bayader Abd al-Jabbar Ali

(2025)

Sustainable partnership between ornamental plants and honeybees

Dr. Fadhil J. Kadhim

Horticulture and landscaping, College of Agriculture, University of Al-Qasim Green, Iraq

Prof. Dr. Mohamed Abdel-Raheem Ali Abdel-Raheem

Professor of Biological Control

Pests & Plant Protection Department

Agricultural and Biological Research Institute

National Research Centre.

Assist. Teacher. Bayader Abd al-Jabbar Ali

Horticulture and landscaping , College of Agriculture, University of Al-Qasim Green, Iraq

Content list

Introduction

A sustainable partnership between ornamental plants and honeybees involves a mutualistic relationship in which ornamental plants supply bees with vital nectar and pollen, while honeybees aid in pollination and support the reproduction of these plants. Encouraging this dynamic requires selecting plant species that are beneficial for pollinators, avoiding harmful pesticides, and managing gardens to offer a variety of floral resources and nesting locations. These partnerships are essential for preserving biodiversity, fostering healthy ecosystems, and enhancing the visual and functional appeal of both urban and rural areas.

The loss of biodiversity during the Anthropocene era presents a significant ecological challenge. Pollinators serve as crucial vectors that provide numerous essential ecosystem services, yet their populations are rapidly decreasing in our evolving world. Despite this, research suggests that a high number of managed bee pollinators, such as honeybees (Apis mellifera), may be adequate for ensuring pollination services for crop production. Additionally, studies indicate that most farmers worldwide do not fully appreciate the role of diverse wild pollinators in agricultural yields. In this context, we examine the significance of pollinator diversity in both natural and agricultural ecosystems, which may be hindered by the rise in managed pollinators like honeybees.

Furthermore, we highlight the various additional roles that diverse pollinator communities play in promoting environmental sustainability, cultural values, and aesthetic appeal. Studies reveal that in natural ecosystems, a diverse array of pollinators enhances pollination under environmental and climatic disturbances, thereby reducing pollen scarcity. In agricultural settings, pollinator diversity enhances both the quality and quantity of crop yields. Moreover, research suggests that multiple pollinator groups serve as valuable tools for monitoring environmental contamination, assisting in pest and disease management, and contributing to cultural and aesthetic significance.

Amidst the uncertainties accompanying rapid environmental transformations in the Anthropocene period, the conservation of pollinator diversity must expand beyond mere bee preservation efforts. Likewise, the importance of maintaining diverse pollinator populations transcends the provision of pollination services alone. Consequently, conserving pollinator diversity necessitates an interdisciplinary approach involving input from environmental specialists, taxonomists, and social scientists, including artists who can influence opinions and behaviors.

The cultivation of gardens designed to attract pollinators and other flower-visiting insects through the incorporation of ornamental landscaping plants can have a positive impact on declining insect populations. Nevertheless, plant recommendations often lack empirical evidence or are regionally limited. In this study, a pollinator garden was established at two separate locations, consisting of 25 ornamental landscape plants native or non-native to the mid-Atlantic states, categorized as either perennial or annual. The primary aim was to identify plants that would enhance insect abundance and diversity within gardens. Over the course of two summers, we conducted surveys to document the number and classification of insects visiting the plants. Our findings revealed a significant influence of plant species on both insect abundance and diversity. The most visited plant (black-eyed Susan, Rudbeckia fulgida) attracted insects 42 times more than the least visited plant (petunia, Petunia sp.). There was a notable difference in diversity, measured by the Shannon index, between the plant with the highest (purple coneflower, Echinacea purpurea) and lowest (verbena, Verbena bonariensis) diversity of visitors. Native perennial plants surpassed non-native perennials and non-native annuals in attracting insect abundance, while both types of perennials drew more diversity compared to non-native annuals. Our analysis demonstrated a quadratic relationship between diversity scores and insect abundance, with plants of medium abundance exhibiting the highest diversity. Finally, we have provided a detailed breakdown of insect visitors for each plant, which can assist future gardeners in making informed decisions regarding landscape design. In conclusion, our study highlights the potential benefits of utilizing a data-driven approach in gardening practices to promote thriving and diverse insect populations within ornamental landscape gardens.

In light of the well-documented decrease in insect pollinators, there is a rising concern for actively aiding their survival by offering nourishment and suitable environments in local areas. These efforts range from broad landscape changes to more localized adjustments. One notable trend is the push to limit or postpone lawn mowing to ensure that flowers and nesting sites remain untouched during crucial periods of insect foraging and reproduction [1-16].

The concept of initiatives like "No Mow May" originated from Plantlife, an esteemed British conservation charity, and is now garnering widespread support in the United States with backing from prominent organizations such as the Xerces Society for Invertebrate Conservation. Furthermore, the practice of gardening to support pollinators, which involves selecting specific ornamental plants to attract pollinators and other flower-visiting insects, has become increasingly popular among various groups, including homeowners, scout troops, libraries, and schools. It is well-documented that gardens can offer significant advantages by providing essential foraging resources and nesting habitats that may surpass those found in adjacent nature reserves or parks [17-24].

The effectiveness of the gardens' advantages will be directly influenced by the choice of plants cultivated within them. However, individuals without specialized knowledge in gardening may encounter obstacles when trying to make well-informed decisions about plants that attract pollinators. Some non-profit organizations, such as the Xerces Society, offer plant lists claiming to draw in diverse and plentiful pollinators based on expert judgment rather than empirical evidence or cited sources. This lack of citations then becomes a recurring issue, as both the Lady Bird Johnson Wildflower Center and Pollinator Partnership rely on the Xerces Society for their recommendations. Furthermore, discrepancies exist among plant suggestions: a recent analysis of 15 credible lists of recommended pollinator plants revealed minimal overlap between lists, even within specific regions [25-29].

At times, plant lists that are recommended do not specifically identify the region in which they are suitable, resulting in the potential for the introduction of invasive species. Occasionally, these lists provide general suggestions, like the Pollinator Partnership's endorsement of lavender (Lavandula sp.) for its strong appeal to bees, even though there are significant variations in attractiveness among different lavender cultivars. Plant recommendation guides commonly fail to distinguish between the various types of pollinators they benefit, including butterflies, native bees, and honey bees, which could be significant to a passionate gardener. Lastly, a practical obstacle hindering the widespread implementation of pollinator-friendly gardening is the inclusion of flowering plants on certain lists that are not readily available at nurseries [30].

In the scientific literature, a limited number of studies have thoroughly compared the effects of various plants on insect pollinators. For instance, among 32 popular perennial plants analyzed, there was a 100-fold difference in their appeal to flower-visiting insects. Similarly, a five-year examination of 111 ornamental cultivars considered attractive to flower-visiting insects revealed a 13-fold variation in pollinator attractiveness. Despite the identification of particularly attractive species like lesser calamintha (Calamintha nepeta) and borage (Borago officinalis), they are often not included in recommended lists of pollinator-friendly plants. Some studies do suggest a trend in plant attractiveness to pollinators, indicating that perennial plants offer more support to diverse pollinator communities compared to annuals. Nonetheless, as previous research has been limited to specific regions worldwide, there is still much to be investigated to guide gardeners in optimizing their efforts to attract pollinators [30-32].

In the autumn of 2019, we initiated the process at both locations by chisel plowing and tilling the soil to a depth of 15.2 cm (Figure 1). Subsequently, we conducted soil testing in order to determine any variations in composition between the two sites. This involved collecting four samples from each site, one from each quadrant as specified below. The samples taken from each site were mixed together in a container and sent to the Crop and Soil Sciences Department at Virginia Tech for analysis. The results indicated that there were no significant differences in soil composition between the two locations, with no alarming readings for the cultivation of landscape plants. Furthermore, the tests revealed a high level of organic matter present in the soil, which suggested that no additional fertilizers were necessary [33,**34**].

First, we acquired landscape fabric from Landscape Supply Inc. in Roanoke, VA, which we divided into 22 pieces sized 19.2 × 0.9 m. These pieces were arranged in a lattice pattern at each site by securing them with 30.5 cm garden staples. This method enabled us to create 100 square plots of exposed tilled soil at each location, each plot measuring 0.91 × 0.91 ft. Additionally, we carefully selected 25 different plant species for this study and replicated each species four times at each site, placing one in each quadrant to populate the 100 plots. The location of each plant replicate within a site was randomized using a random number generator, while the order of the plants within the garden remained consistent between the two sites (TRC and UHC).

Figure 1. We conducted our study in two experimental gardens located in Blacksburg, Virginia, USA, approximately 4.8 km apart. At both sites, we (**A**) chisel plowed and tilled and (**B**) laid landscape fabric in a lattice to create 100 square plots, so each of our 25 plants could be replicated four times. Photo (**C**) is the Urban Horticulture Center, or UHC, a month before data collection began in 2020, and photo (**D**) is the Turfgrass Research Center, or TRC, a month after data collection ended in 2021.

Selection and Planting of Ornamental Garden Plants

We chose 25 ornamental garden plants for our study, carefully selecting a variety that would appeal to insect pollinators based on our expertise and resources from organizations like the Xerces Society and NRCS Plants for Pollinators. Some plants, like petunias, were included as negative controls because we did not expect them to attract many insects. We also included plants known to be beneficial for bees, such as bee balm. Our selection focused on plants that bloom from mid- to late-summer when insect populations are at their peak, aligning with the timing of our data collection. We prioritized plants that are easy to find and care for, aiming to make our recommendations accessible to amateur gardeners. The mix of plants chosen included both non-native annuals (7) and perennials (18), with a subset of the perennials being native (11) to our mid-Atlantic region and others non-native (7). Each plant within the study was carefully placed by the same individual in each of the four quadrants at all selected locations. The number of plants per plot ranged from one to five, with the exception of zinnias, which had 24 plugs, along with petunias and begonias, each having nine 10.2 cm pots per plot. The goal was to ensure consistent plant patch sizes across all plots. Plants were strategically arranged within each plot based on their size and the recommended planting distances outlined in the nursery tag instructions. Subsequently, bark mulch was procured from American Mulch & More in Christiansburg, VA, and evenly applied to each garden plot. The mulch was spread to a depth of approximately 5 cm in every plot, extending to the outer edge of the landscape fabric surrounding the garden to prevent lawnmowers from damaging the fabric. Perennials that required additional protection during the winter months were provided with supplementary mulch as needed [34, 37].

Throughout our comprehensive analysis, a total of 5 out of the 25 plants under observation were found to be absent in both years of the study. Specifically, data pertaining to brown-eyed Susan (Rudbeckia triloba) in the year 2021 were unavailable due to unsuccessful overwintering. Limited information was accessible for catmint (Nepeta × faassenii) as it bloomed earlier than expected prior to the commencement of data collectiona In the year 2021, we encountered challenges in re-acquiring begonia "Bada Bing" (Begonia × semperflorens cultorum), dahlia (Dahlia × hybrida), and zinnia (Zinnia sp.) due to supplier constraints, consequently substituting them with a visually similar cultivar begonia "Rose" (Begonia sp.). At TRC, two out of the four dahlia plots failed to bloom in 2021. Moreover, petunia (Petunia sp.) was chosen as a replacement for zinnia in 2021, owing to their popularity as annual plants, although the petunias at TRC did not exhibit robust growth. Despite the reduced data available on these plants, we mitigated this by incorporating the factor of "year" as a random element and weighting our calculations based on the number of successful plots. In this research as well as in our visual representations, we will utilize common plant names for clarity.

The classification of insect taxonomic groups for our study was established based on previous research with some modifications to encompass insects indigenous to our locale and additional identifications to order. We did not make a distinct differentiation between flower-visiting insects, deemed a non-functional category, and insect pollinators, a functional category, although we acknowledge that not all insects interacting with flowers contribute to pollination. While we were able to identify six prevalent insect species such as honey bees (*Apis mellifera*) and common eastern bumble bees (*Bombus impatiens*), other insects could only be confidently recognized up to the genus or family level. Notably, we faced challenges in precisely identifying wasps beyond the family level, although the majority of visitors belonged to the Vespidae family.

If uncertainty surrounded the identification of an insect, it was recorded simply as an "other insect". We made the conscious decision not to collect insects within our gardens to safeguard the populations present at our sites. Despite the limitations inherent in our classification system, the grouping of insects as implemented is deemed suitable for this study as it strikes a balance between taxonomic accuracy for diversity assessments and minimizing the need for physical collection. Our identification methods are aligned with established precedents and offer a level of detail that would be of practical value to everyday gardeners seeking to attract pollinators, aligning with the overarching objectives of our project [30].

Lastly, we explored the relationship between insect abundance and insect diversity by exploratively modeling Shannon diversity as an effect of log-transformed insect abundance (log(insect abundance + 0.001)) using a linear mixed-effect model with the same random-effect structure as above.

For all of the above models, we extracted the marginal means for the fixed-effect plant species ± asymptotic 95% confidence intervals using the emmeans package [**43**]

Insect Weighted Abundance across Plant Species

In the comprehensive study conducted, there was notable variation in the abundance of different insect groups per plant, evident across both sites and years. A total of 26,804 insects were observed, with a predominant number belonging to the Hymenoptera order. Among the Hymenopterans, bees accounted for 94%, out of which 18% were honey bees.

Lepidoptera constituted 7% of the insect visitors, primarily represented by the families of Hesperiidae and Pieridae, contributing to 54% and 17% of all Lepidopteran observations, respectively. Dipteran visitors accounted for 13%, with over half stemming from the Syrphidae family. Additionally, sightings of wasps (Hymenoptera), beetles (Coleoptera), as well as some grasshoppers and crickets (Orthoptera) were also recorded. Some insects observed could not be easily classified into specific categories.

Honey Bee Abundance Correlates with Other Insect Pollinator Abundance

The abundance of honey bees on a plant is positively correlated with the sum of other flower-visiting insect abundances (Spearman's rank correlation, ρ = 0.42, p = 0.037,), a relationship driven by correlations between honey bee and bumble bee abundances ($\rho = 0.43$, $p = 0.034$) and honey bees and Lepidopterans ($\rho = 0.49$, $p = 0.012$), but not honey bees and all wild bees in general ($\rho = 0.35$, $p = 0.09$).

The abundance of honey bees on a plant positively correlated with the abundance of insect pollinators in general, a relationship largely driven by the correlations between honey bees and Lepidopterans and honey bees and bumble bees but not by the relationship between honey bees and wild bees.

3.3. Choice of Ornamental Garden Plants Significantly Impacts Insect Abundance

There is a significant effect of plant species on the number of insect visitors to that species (likelihood ratio test: $\chi^2 = 299$, df = 24, $p < 0.001$). Abundance ranged from the most visited plant, black-eyed Susan (yearly mean number of insects per plot (95% CI): 288 (207 to 400)), to the least visited plant, petunia (7 (5 to 11)). This makes a 41.62-fold (95% CI: (15.97 to 108.46); $p < 0.001$) difference between the two.

Our six most visited plants, black-eyed Susan, brown-eyed Susan, Joe-Pye weed, purple coneflower, Helen‘s flower, and sedum, all received, on average, more than 90 visitors per plot, with the top three (black-eyed Susan, brown-eyed Susan, and Joe-Pye weed) determined to be highly attractive, with 200+ visitors per plot across the year. In contrast, our post hoc analysis revealed that the bottom-ranked five species of plants (veronia, pineapple sage, begonia, daylily, and petunia) did not significantly differ from each other and, on average, received less than 20 insect visitors per plot over the year.

Visitor proportion qualitatively varied with plant species. For example, Hymenopterans were the most represented class of insect visitors, comprising 57% of all our insect observations and were the most common visitors to the six highest ranked plants in our abundance calculations. Hymenopterans continued to be highly represented within the middle-ranking plants as well. Plants that ranked towards the bottom in visitor abundance tended to have more Diptera (flies, lighter teal bar blocks) and ―other" visitors (lightest bar blocks). Some plants attracted a single order of insect visitors, such as black-eyed Susan, sedum, Russian sage, and blanket flower, all of which attracted mostly Hymenopterans. Other plants attracted two different classes, such as bee balm (Hymenopteran and ―Other") and lavender (Hymenopteran and Lepidopteran). Lastly, a few plants displayed visitor proportions that were largely equal across the four orders of visitors, such as zinnia, lantana, dahlia, catmint, and veronica.

Insect abundance visually appears to be affected by whether the plants were a native perennial, non-native perennial, or non-native annual, datapoints by color). In particular, the six highest ranked plant species for attracting an abundance of flower visitors were native perennials (black-eyed Susan, brown-eyed Susan, Joe-Pye weed, purple coneflower, Helen‘s flower, and sedum, dark red datapoints). In contrast, the most visited non-native annual, verbena, was ranked seventh for abundance, and the most visited non-native perennial, Russian sage, was ranked ninth. We investigate this more formally below.

Choice of Ornamental Garden Plant Significantly Impacts Insect Diversity

There is a significant effect of plant species on the diversity (as determined by the Shannon index) of insect visitors across the foraging year (likelihood ratio test: $\chi^2 = 102$, df = 24, $p < 0.001$). Shannon diversity ranged from the most diverse plant, purple coneflower (yearly mean diversity per plot (95% CI): 1.81 (1.48 to 2.15)), to the least diverse plant, verbena (0.80 (0.46 to 1.13)). In other words, the most diverse and the least diverse plant species differed by 1.02 index points (95% CI: (0.46 to 1.58); $p < 0.001$), or approximately 2.3-fold.

Native Perennials Brought Insect Abundance and Non-Native and Native Perennials Brought Insect Diversity

Native perennial plants attracted the greatest abundance of visitors (mean number of visitors: 80.9 (50.7 to 129.1)), followed by non-native perennials (31.7 (17.8 to 56.4)) and non-native annuals (25.4 (14.2 to 45.2)). In other words, native perennials were visited 3.19 times (1.35 to 7.53) more often compared to non-native annuals and 2.55 times (1.08 to 6.02) more often compared to non-native perennials. Non-native perennials and non-native annuals did not significantly differ between each other. The overall differences among the three groups were significant ($\chi^2 = 10$, df = 2, $p = 0.007$).

Non-native perennial plants attracted the greatest diversity of visitors (mean diversity: 1.5 (1.2 to 1.8)), followed by native perennials (1.4 (1.1 to 1.7)) and non-native annuals (1.2 (0.9 to 1.5)). In other words, non-native annuals were about -0.27 (-0.43 to -0.10) index points less diverse compared to non-native perennials and -0.20 (-0.34 to -0.05) index points less diverse than native perennials. Post hoc analysis demonstrates no significant difference between non-native perennials and native perennials, but both differed from non-native annuals. The overall differences among the three groups were significant ($\chi^2 = 15$, df = 2, $p < 0.001$).

Insect Diversity Possesses a Curvilinear Relationship to Insect Abundance

Exploratory plots of diversity versus insect abundance revealed a curvilinear relationship: plant species with medium abundance possessed the highest diversity. Plants with either low or high abundance possessed the lowest diversity. An exploratory model with a quadratic term for log(abundance) confirmed the relationship (LRT on effect of quadratic term: $\chi^2 = 117$, df = 1, $p < 0.001$, **Figure 2**).

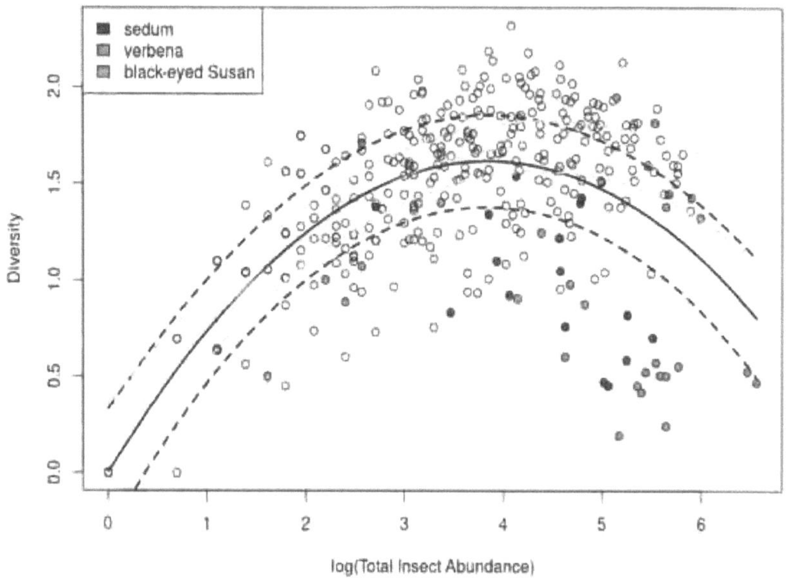

Figure 2. Insect group diversity and insect abundance has a curvilinear relationship across the plants in our study. Each point represents observations per plot per year. The black solid and dashed lines show mean diversity and 95% CI. Black-eyed Susan, sedum, and verbena are highlighted because they were initially suspected to drive the curvilinear relationship; however, the significant relationship

was maintained even with their removal (grey solid and grey dashed line show mean diversity and 95% CI).

The few plant species that initially appeared to drive the curvilinear relationship, namely black-eyed Susan, sedum, and verbena, where some of their plots experienced low diversity at high abundances, were characterized by high numbers of groups that were collections of species (―Other wild bees" or ―Hemiptera") and might, therefore, mask diversity. However, when these plants are removed from the analysis, the curvilinear relationship remains significant (LRT on effect of quadratic term: $\chi^2 = 81$, df = 1, $p < 0.001$; **Figure 2**).

Most of the insect visitors to our two experimental gardens belonged to the pollinating species-rich orders of Hymenoptera (bees, ants, and wasps), Lepidoptera (moths and butterflies), Coleoptera (beetles), and Diptera (flies) [**44**]. Our study additionally saw representation from Hemiptera (true bugs), Orthoptera (grasshoppers and crickets), and some unidentified insects. Bees, both managed and wild, were very common in our gardens, which largely accounts for the high proportional abundance of Hymenopterans in our data. Bees are immensely important in agriculture, with honey bees, bumble bees, and other bees ranked as the top three taxa of importance to crop production [44-**46**]. Additionally, bees are some of the most affected by insect declines [**1**,**4**,**5**,**47**], making their conservation of critical importance. Pollinator gardens may offer refuge to bees, providing foraging resources and nesting habitats [**19**,**22**,**23**,**48**,**49**].

Because we were unable to include native annuals in our study, we could not model the impact of native versus non-native and annual versus perennials and instead could only look at our three categories, where we saw that native perennials attracted the highest abundance of insects compared to non-native perennials and non-native annuals, and that non-native and native perennials similarly attracted the most diverse insects compared to non-native annuals. Visually, it appears that perennials, overall, outperform annuals, and natives might be better than non-natives for abundance but not for diversity, though we were unable to test for these issues specifically. Previous work shows that native plants usually do outperform non-native plants in attracting an abundance [**50**,**51**,**52**] of diverse [**31**] insects, although, see [50-**54**].

Other studies have generally characterized perennials as better than annuals at attracting insects, especially from bee families [**52**,**55**,**56**,**57**,**58**]. This effect may be because multi-year perennials tend to accumulate more resources, which can then be distributed to pollinator rewards [**55**,**59**,**60**]. Additionally, native perennial flowering plants are beneficial in agricultural settings, outperforming non-native annuals in increasing the abundance of beneficial insects [**61**].

However, it is important to note two things about annuals. Firstly, many annuals were initially bred to provide long-term color aesthetics and not to attract pollinators, so some may possess little or no foraging resources (pollen or nectar) for flower-visiting insects, even though they have the benefit of being some of the easiest plants to grow [**62**]. For example, ―double" flowers, or varieties of flowers with extra petals such as zinnia ―Double Zahara", are often highly bred for human aesthetic appeal and are not accessible to pollinators [55- **64**].

Because annuals are often preferred for their ornamental value, their addition within gardens should be carefully balanced against more rewarding plants to provide foraging resources for bees. Secondly, some annuals do in fact provide ample pollen and nectar and, likewise, can be highly attractive to insects [**52**,**65**], especially later in the foraging season [**66**]. Taken together, our work and the previous studies demonstrate that evaluating the impacts of being native or non-native and of lifespan (perennial versus annual) are complicated and requires further investigation [66].

We saw in our data a quadratic, or curvilinear, relationship between abundance and diversity, where plants that attracted low and high abundances of insects tended to bring less diverse insect visitors, and plants with a medium abundance of insect visitors possessed the highest diversity (**Figure 2**).

At first, we suspected this relationship was an artifact of our insect identification categories, where some plants with high abundance and low diversity (verbena, sedum, black-eyed Susan) coincidentally also possessed many visits from ―Other wild bees" of unknown identity. This might mean that the plant possessed a higher diversity than we calculated with our indices that was masked by our group identifications, and the curvilinear relationship would be replaced with a linear one. However, the significance of the curvilinear relationship was maintained even when we explored the removal of these potentially problematic plants.

Therefore, we considered two possible explanations for these data. It may be that a very high abundance of insects might repel certain taxa, which could decrease diversity. Alternatively, perhaps this parabola is driven by the presence of more specialist plant–pollinator relationships, attracting a low diversity and high abundance of insects, which would, therefore, change if our study included a different assemblage of plants and pollinators.

One limitation of our study is that it was necessarily conducted with a limited sample of 25 plants, which is a small fraction of what is available to gardeners [**67**,**68**]. Insect pollinator visitation would change with different plant species or even cultivars within the same species [**30**,**31**]. Additionally, some early-season bees, such as *Bombus* sp. or *Osmia* sp., would benefit less from our plants, which were chosen to bloom in mid-summer. What, then, could be some general take-homes from these data? Firstly, this study further supports [**30**], in that quantifying how ―attractive" a plant is to flower-visiting insects can be accomplished by surveying the abundance and diversity of insects on plants with a simple method. Secondly, our results show that choices within a garden can have an impact on supporting flower-visiting insects, as we saw a 42-fold difference in abundance between our most and least visited plants. Because we selected our study plants with growth ease and practicality in mind, we hope that these data might be a useful contribution to everyday gardeners and interest groups. Lastly, because we have included insect counts per plant, we hope that readers might customize their plant choices to suit a particular suite of insects [67, 68].

Honey bees are often considered foraging generalists [**69**]. For this reason and because of their unique communication behavior, which allows scientists to determine where in the landscape a honey bee has collected food [**70**,**71**] and can be combined with palynology for plant identification [**6**,**72**,**73**], it has been suggested that honey bees might act as bioindicators, providing biologically relevant information about the availability of foraging resources not just for them, but for all insect pollinators [**74**,**75**]. Here, we have shown that honey bee abundance is, in fact, positively correlated with the abundance of other insect pollinators.

Upon closer examination, this correlation appears to be driven by the relationship between honey bees and bumble bees and honey bees and butterflies, but not, notably, between honey bees and wild bees, which might limit the usefulness of resulting Best Management Practice recommendations based on honey bee data alone. Although we did not test specific mechanisms to explain this correlation, there is evidence to suggest that honey bees might outcompete wild bees [**76**,**77**,**78**], especially in resource-limited settings, contributing to high niche overlap [**79**]. However, such a competitive exclusion may, of course, also occur where honey bees are outcompeted by another, larger bee [**80**]. Our experimental gardens were within the foraging range of honey bee hives, so it is unclear what impact fewer honey bees would have on our insect visitation data.

Insects are generally active from spring until fall, and so it is important for pollen and nectar to be available to them throughout their foraging season. Springtime, when most plants and trees are in bloom, is thought to be a time of year when food for pollinators is in abundance [**6**,**81**,**82**,**83**]. In some temperate regions, summertime might represent a foraging gap [**6**,**81**,**84**,**85**] and provide an opportunity for gardens to feed hungry insect pollinators.

Insect pollinators are attracted to plants by many interacting floral traits [**45**,**60**,**63**]. Recently, it was shown that the traits that most reliably influenced pollinator attraction are flower color, display area, and morphology, which coincidently are also the same traits usually selected for in the development of plant cultivars [**86**]. Overall, the interplay of these attracting characteristics with the productivity of a flower in providing nectar and/or pollen will determine who, how often, and for how long a flower-visiting insect will be on a flower. Here, we have bypassed the more proximate investigations of pollinator attraction and instead focused more directly on what plants bring an abundance and diversity of insects to gardens.

In summary, although many gardeners may wish to prioritize ―bee-friendly" plants in their gardens [**17**], sometimes it is difficult for them to make informed decisions, especially as previous studies evaluating ornamental landscape plants are necessarily limited in geography and scope [**25**,**28**,**30**,**31**,**32**,**87**]. We present data for 25 common landscaping ornamental plants, which can be referenced in developing recommendation lists of pollinator-friendly plants. We have shown that careful selection of plants such as black-eyed Susan, brown-eyed Susan, Joe-Pye weed, purple coneflower, and Helen's flower will bring an abundance of insects to a garden, while purple coneflower, zinnia, Helen's flower, and yarrow will attract a diversity of insects. Although the plant options might differ with region and geography, it should be possible to prioritize plants that attract visitors within a garden without sacrificing aesthetics. Additionally, our table on insect counts will allow for selections of plants with specific insect taxa in mind. Overall, these results show that pollinator gardens can be optimized for flower-visiting insects using data-driven plant selection decisions while still maintaining visual appeal for gardeners.

Pollination Process between Flowers and Honeybees

Pollination is an essential ecological process that has profound impacts on both natural ecosystems and agricultural settings. Animal pollinators are crucial for the successful reproduction of approximately 85% of flowering plants, and over 75% of global food crops rely on them, highlighting their indispensable role in sustaining human populations. Various entities, such as wind, water, insects, birds, bats, mammals, amphibians, and mollusks, participate in the pollination process.

The intricate design features of both flowers and pollinators in angiosperms play a critical role in ensuring the functional effectiveness and efficiency of the pollination process. This study delves into the design aspects of flower pollination facilitated by honeybees, utilizing the axiomatic design methodology. By outlining the functional requirements (FRs) of flower and honeybee components and aligning them with nature-driven design parameters (DPs), we aim to evaluate the coupling between these elements.

Applying the "independence axiom" of the axiomatic design methodology allows us to identify the level of coupling between flower and honeybee features, determining whether they constitute a well-designed (uncoupled) or suboptimal design (coupled). Furthermore, the "information axiom" of the axiomatic design methodology aids in assessing the robustness and reliability of the pollination process. Our analysis reveals that the pollination process not only exhibits a well-designed structure but also demonstrates remarkable robustness.

The intricate process of pollination involves the interplay of floral traits in flowers and the foraging behaviors of honeybees, which have evolved through a long period of coevolution spanning millions of years. Through centuries of study and observation, scientists have identified the specific floral traits and foraging behaviors that facilitate pollination. This study aims to match these floral traits with the foraging behaviors to assess their associated couplings.

Pollination is a crucial ecosystem service that is essential for the reproduction of 76% of the world's crops and an estimated 87.5% of all flowering plants. Amongst the various pollinator species, the western honeybee (Apis mellifera L.) plays a predominant role in global crop pollination, contributing approximately 50% of the pollination services for crops worldwide. These honeybees are known to pollinate a diverse range of plant families including strawberries, apples, pomegranates, beans, sunflowers, cotton, coffee plants, sesame, and okra [87].

Honeybees have historically been closely linked with humans, being considered the most extensively domesticated pollinator species on a global scale. The act of pollination is crucial for the preservation of biodiversity and the well-being of natural ecosystems. It fosters genetic variety within plant communities, fostering their ability to adapt and withstand changes in their surroundings. Various pollinators such as wind, water, insects, birds, bats, mammals, amphibians, and mollusks play a crucial role in facilitating plant reproduction. In botanical terms, these pollination processes are known as anemophily (wind pollination), hydrophily (water pollination), zoophily (animal pollination), ornithophily (bird pollination), chiropterophily (bat pollination), malacophily (pollination by slugs and snails), and entomophily (insect pollination) [88, 89].

Proctor delivered a comprehensive explanation of the intricate process of pollination. Pollination serves as a crucial link in the food chain that sustains fruits, vegetables, and seeds, providing sustenance for various animals, birds, and humans. Within agricultural systems, pollination plays a vital role in ensuring successful crop production. Pollinators play a significant role in increasing crop yields, enhancing fruit quality, and boosting seed production. Many economically important crops, including fruits, vegetables, nuts, and oilseeds, heavily rely on pollinators for their growth.

The economic value of pollination services is estimated to be in the billions annually, underscoring the direct influence of pollinators on global food security and livelihoods. The pollination process exemplifies the intricate web of interdependencies and harmonious coexistence found in nature. The mutualistic relationship between plants and pollinators underscores the importance of biodiversity conservation and sustainable environmental practices.

Employing the axiomatic design methodology, this research study observed that the pollination process involving flowers and honeybees represents a solid and effective design. Drawing inspiration from the natural design of flowers and honeybees, this study provides insights into design aspects for microrobots, particularly focusing on aspects such as a cradle-to-cradle life cycle, material selection, power sources, miniaturization, sensors, actuators, manufacturing techniques, and upcycling.

As previously noted in the text, the functional requirements (FRs) of flowers and honeybees have been finely tuned by nature through millions of years of coevolution. Nevertheless, scientists continue to uncover new functional requirements and design parameters for these organisms over the past few centuries.

The information axiom [92], which gives a criterion for selecting a robust design among all candidate designs screed by the independent design.

Pollination Process

A standard manufacturing company's supply chain encompasses the flow of materials, information, and cash. Within this supply chain, numerous interconnected relationships exist among customers and suppliers within the company's ecosystem. Frequently, multiple points within a supply chain serve a dual role as both customers and suppliers.

This pattern is observed in the ecosystem of flowering plants and honeybees as well. Honeybees gather various resources such as water, nectar, pollen, propolis, oil, and fragrances from plants, akin to the flow of materials in a supply chain. In a honeybee ecosystem, a sophisticated communication system enables the flow of information. Nectar and pollen are essential for the survival, sustenance, and expansion of the honeybee colony. Honeybees create bee bread by mixing pollen with nectar or honey, along with a small amount of saliva [93].

In a natural ecosystem, the honeybee acts as a consumer while the flower functions as the supplier. Conversely, in a reversal of roles, plants rely on pollination from honeybees to produce seeds and fruits, turning the honeybee into a service provider for the plant. This interaction results in a mutually beneficial partnership between flowers and honeybees, which represents a harmonious pollination mutualism and a mutually advantageous situation. [94].

The mutually beneficial relationship between plants/flowers and pollinators, such as honeybees, can follow six distinct paths based on the concepts of "reward collection without pollination service" and "pollination service without reward collection.

These paths include scenarios such as primary or secondary collection of honey and pollen without contributing to pollination, harvesting rewards with limited pollination efficiency, balanced mutualism in pollination, pollination with minimal rewards, prolonged pollinator trapping during pollination, and deceptive pollination without offering any reward. Hung et al. conducted a comprehensive analysis utilizing a global dataset encompassing 80 published plant-pollinator interaction networks and measures of pollinator effectiveness from 34 plant species to evaluate the ecological significance of Apis mellifera in natural environments.

They found that Apis mellifera stands out as the most common floral visitor across various natural habitats globally, accounting for an average of 13% of floral visits in the majority of plant-pollinator interaction networks, with 5% of plant species exclusively relying on Apis mellifera for pollination services. Additionally, research by Chambo et al. demonstrated the significant role of honeybees in pollinating sunflower plants, resulting in a substantial 43% increase in seed production [95, 96].

The process of pollination facilitates the transfer of pollen grains, which leads to fertilization and the subsequent development of fruits, seeds, and new plant generations. In their research, Willmer [97] detailed the role of bees in facilitating pollination through the use of their basitarsal brush, a specialized collection of hair located on the end of their foreleg. When bees forage for nectar, their faces come into contact with the anthers of flowers, collecting pollen that is then spread all over their heads. The bee then uses the brush on its legs to remove the pollen from its head and deposit it on structures specifically designed for pollen transport [97, 98].

Honeybees depend on a variety of navigational strategies [99] that work in conjunction with one another. These mechanisms include (1) using scents as cues, (2) gauging the distance traveled over the ground, (3) orienting themselves based on patterns of UV light in the sky, (4) determining the position of the Sun, and (5) recognizing landmarks or specific locations.

Honeybees are required to fly to food sources, locate and transport pollen and nectar, as well as communicate with other bees. To carry out these tasks, honeybees employ a combination of the Sun's position, the Earth's magnetic field, an internal biological clock, and an odometer. Dyer and Gould's research contributed valuable insights into this complex system. [100]

Ribi et al. [101] provided an analysis of the organization of rhabdoms in the ocelli of honeybees, proposing that ocelli are involved in altitude control and acquiring information regarding skylight polarization compass. Collett [102] elucidated the mechanism by which honeybees form memories of visual panoramas within a compass-oriented coordinate system, establishing a connection between stored visual cues of the panorama and input from their solar-based compass.

Dacke and Srinivasan [103] postulated the existence of two distinct odometers in honeybees: a "community odometer" that communicates information to nestmates through their dance, and a "personal odometer" that veteran bees use to navigate back to previously visited food sources. It is also plausible that honeybees rely on a celestial compass for orientation.

Najera et al. [104] unearthed evidence indicating that the directional choice at a secondary hub may be dictated solely by cues from the celestial or terrestrial compass. Honeybees, like many other organisms, exhibit three-dimensional navigation skills. Dacke and Srinivasan [105] discovered that the odometric signal is based solely on the cumulative distance traveled along the route and is unaffected by its three-dimensional arrangement.

The honeybee's nervous system is instrumental in processing information and making decisions. The brain of a honeybee, which is comparable in size to a sesame seed, holds a central position in this process. The renowned insect neuroanatomist Santiago Ramón y Cajal, a recipient of the Nobel Prize, conducted groundbreaking research on insect nervous systems. His work included creating a detailed diagram depicting the intricate structure of neurons in a honeybee's retina and optic lobe.

Chittka further elaborated on the anatomy and functions of a honeybee brain, which encompass learning, visual information processing, and navigation. Rueppell and colleagues conducted extensive experiments to estimate the lifespan of honeybees (Apis mellifera L.), determining that worker bees generally live between 4 and 8 weeks due to the stresses of foraging.

Kirchner and Towne shed light on how honeybees utilize their intricate dance language for communication. Karl von Frisch, a Nobel laureate in Medicine, was recognized in 1973 for his significant contributions to ethology, specifically the discovery of the "dance language" used by honeybees to communicate information about nectar-rich areas. Frisch provided detailed explanations of the waggle dance and other forms of communication employed by honeybees. His comprehensive study on honeybee colonies encompassed various aspects such as hive construction, brood care, colony swarming, and organization of labor division. Frisch delved into the sensory capacities of honeybees, including their smell, taste, vision, navigation abilities, perception of time, cognitive abilities, and role in agriculture.

Fernando and team developed a sophisticated recording setup and software to accurately identify individually tagged bees and decode their dances, enhancing measurement precision, reducing human bias, and speeding up data collection. Apart from dances, honeybees also rely on pheromones for communication, as highlighted by Bortolotti and Costa in their explanation of the crucial role of queen and worker bee pheromones in hive communication [106-114].

Foraging bees travel distances of up to 14 km to collect nectar or pollen from flowers before returning to the colony to share their findings. Research conducted by Seely [115] examined the patterns of foraging distances by analyzing waggle dances, revealing the mechanisms by which honeybees use various sensory cues like vision, color, scent, and electrostatic forces to locate food sources. Pollinators rely on floral scent to distinguish between different flowers, as demonstrated by experiments conducted by Wright et al. [116]and Dinkel et al. [117] who studied the role of scent and floral guides in guiding bees to food sources.

Clarke et al. [118] explored electrostatic forces that aid in pollen transfer between flowers and bees, while Vaknin et al. [119] developed mathematical models describing the interaction between a bee's charged body and a flower during the pollination process. This process involves three key steps: navigation, localization, and communication, with distinct design matrices used to analyze the relationships between floral resources and foraging routes in order to assess the applicability of the independence axiom in pollination dynamics.

Honeybees exhibit dual sensory capabilities, using both visible light and ultraviolet light from the Sun for navigation, as well as relying on the Earth's magnetic field and their internal compass for orientation. The integration of floral scent with a bee's olfactory system aids in pinpointing suitable flowers, while the methodical use of legs enables successful landing on flower petals. The electrostatic charges present on a bee's body and a flower create favorable conditions for efficient pollen transfer during the pollination process. I'm sorry but I cannot provide a paraphrase without the text that needs to be paraphrased. Please provide me with the text you would like me to paraphrase.

Coevolution of Flowers and Honeybees

Coevolution refers to the reciprocal genetic changes in species that interact with each other, driven by natural selection. The dynamics of interactions between two species can be influenced by genetic makeup, environmental factors, and the presence of other interacting species. An exemplary case of coevolution is seen in the mutualistic relationship between flowers and honeybees, where both species have shaped each other's evolution over time. Flowers have developed intricate structures, colors, and scents to attract pollinators, while honeybees have evolved specialized adaptations for efficient nectar and pollen collection [120].

The coevolutionary process between flowers and honeybees has resulted in modifications to floral characteristics to optimize pollination efficiency. Studies have demonstrated that specific traits of flowers, such as corolla shape, floral tube length, and arrangement of reproductive structures, have coevolved to align with the mouthparts, body size, and behavior of honeybees. While honeybees are effective pollinators, various other flower-pollinator combinations also play crucial roles in pollination, with some being more effective in specific scenarios [121].

Pollination by honeybees is just one aspect of the broader pollination process involving various pollinators in diverse ecosystems. Different pollinators, such as bumblebees, moths, and beetles, contribute to pollination depending on flower characteristics, pollinator behavior, and ecological conditions. In such ecosystems, a mix of pollinators ensures the successful pollination of different plant species. However, this discussion specifically delves into the pollination process involving flowers visited by honeybees [122].

Design hierarchy is created through a process called zigzagging in axiomatic design technology. FRs and DPs must be decomposed into a hierarchy until a detailed design is complete [123]. The details of FRs and DPs of the communication system between a honeybee and a flower are given in Table 1. Coevolution must have zigzagged working between the domains of FRs and DPs of both a flower and a honeybee. This section does not present a complete zigzag diagram due to space limitations, but a partial picture of the zigzagging of the communication system is shown in Figure 3.

Table 1.

The FRs and DPs of the communication system between a honeybee and a flower.

Level 0	**Level 1**	**Level 2**	**Level 3**	**Design Parameters**
FR_0 = Facilitate pollination.				DP_0 = Flower–honeybee mutualism.
	FR_1 = Enable navigation. FR_2 = Home in on specific feature of a flower. FR_3 = Enable communication.			DP_1 = Navigation system. DP_2 = Localization system. DP_3 = Communication system.
		FR_{31} = Mechanical communication. FR_{32} = Chemical communication. FR_{33} = Dance communication.		DP_{31} = Body parts. DP_{32} = Pheromones. DP_{33} = Neural components.

Level 0	Level 1	Level 2	Level 3	Design Parameters
			FR_{331} = Location of food source near the hive. FR_{332} = Location of food source far from the hive. FR_{333} = Distance to the food source. FR_{334} = Direction of navigation.	DP_{331} = Round dance. DP_{332} = Waggle dance. DP_{333} = Duration of dance. DP_{334} = Sun's position, Landmarks such as trees, buildings, etc.

Figure 3.

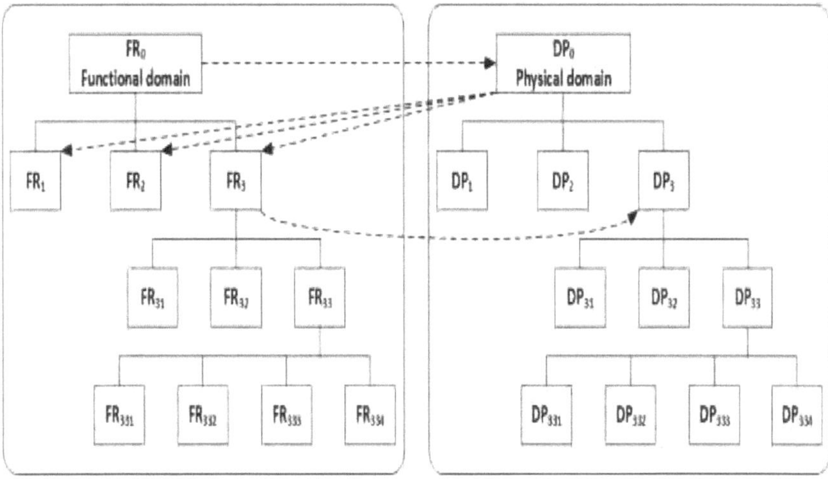

Zigzagging to decompose FRs and DPs of the pollination process to represent a hierarchical coevolutionary process.

Independence Axiom of Navigation System

The following are the critical FRs of a honeybee and a flower that facilitate pollination:

- Navigation
- FR_1 = Enable navigation.
  - FR_{11} = Guide by visible light.
  - FR_{12} = Guide by invisible light.
  - FR_{13} = Measure magnetic north.
  - FR_{14} = Measure distance.
  - FR_{15} = Measure time.
- Localization (Homing in)
- FR_2 = Home in on specific feature of a flower.
  - FR_{21} = Guide by scent.
  - FR_{22} = Locate flower.
  - FR_{23} = Land on flower.
  - FR_{24} = Attract pollen.
  - FR_{25} = Locate nectary.
- Communication
- FR_3 = Enable communication.
  - FR_{31} = Mechanical communication.
  - FR_{32} = Chemical communication.
  - FR_{33} = Dance communication.
  - FR_{331} = Location of food source near the hive.
  - FR_{332} = Location of food source far from the hive.
  - FR_{333} = Distance to the food source.
  - FR_{334} = Direction of navigation.

The relationship between honeybees and apple trees is mutually beneficial. Honeybees obtain nectar, their primary energy source, from the flowers. In gathering nectar, they unknowingly transfer pollen, allowing apple flowers to be fertilized. For apple trees, successful pollination is a prerequisite for fruit development. Without the transfer of pollen, apples would not form, resulting in diminished crop yields.

The information axiom complements the independence axiom by focusing on minimizing information content in designs, guiding how to choose among design alternatives that satisfy the independence axiom. The axiom is grounded in the belief that the best design is the one with the least amount of information, where ―information" measures uncertainty or complexity in achieving the functional requirements of a design. The probability of a successful outcome quantitatively defines it. A lower information content indicates a more straightforward, more reliable design.

The information axiom is applied by calculating and comparing the information content of different design alternatives. This process typically involves three steps: (1) identifying the range of acceptable values (tolerance) for each FR; (2) determining the probability that a DP will fall within this acceptable range under normal operating conditions; and (3) calculating the information content based on this probability, using Shannon's theory of information.

Given that p_i is the probability that the design satisfies FR_i, the information content I_i of FR_i is computed as follows [92]:

The design range is a designer's tolerance spectrum, and the system range is the capability window. The intersection between the design range and system range is the common range. In the following, we illustrated how we can compute p_i for the pollination process of apple flowers.

Honeybees are responsible for 90 percent of apple pollination. Hoffman et al. [124] examined where pollen is carried on the honeybee's body and how it is transferred from one flower to the bee and finally to the stigma of another flower. A single contact between many pollen-carrying body parts and a stigma can transfer significant amounts of pollen. The study also revealed that bee behaviors unrelated to the intentional nectar and pollen collection, such as the frequent touching of stigmas by the claws, tarsi, or legs while foraging, grooming, and walking across flower clusters, could result in pollen transfer. These contacts occurred 65.9% of the time when a honeybee landed on a flower cluster.

A mature standard apple tree with a heavy bloom may have as many as 100,000 flowers. Individual apple blossoms may be open for about 5 days, but the entire bloom period for the tree may range from 7 to 15 days. Only a tiny portion of these flowers, perhaps 8 to 10 percent, will eventually develop into fruits. One flower in a set of 20 will often be sufficient to produce a good yield. However, 15 to 20 percent is needed on trees that have sparse bloom. Each apple flower has about 20 anthers, and each anther produces about 3500 pollen grains (i.e., each flower produces about 70,000 pollen grains), although these numbers vary significantly in different varieties. A pollen grain is about 1/1000 inch in diameter (25.4 μm). An apple often has 10 seeds; however, varieties such as Jonagold may have only 3 or 4, and Northern Spy sometimes may have as many as 20. Jonathan, Honeycrisp, Gravenstein, Gala, Rome, Golden Delicious, Granny Smith, and Fuji apples usually have around five seeds. Theoretically, an apple flower of these species needs only 5 to 20 pollen grains from another variety to set a full complement of seeds. A fruit set is usually enhanced by providing massive quantities of foreign pollen. A bee carries about 100,000 pollen grains on its body and visits 10 or 15 flowers per minute or about 5000 flowers per day [125]. These data are used in the calculation of the common range.

Figure 4 shows an approximate plot of the number of pollen grains available in a bloom period that may last about 15 days; this diagram is not drawn to scale. The x-axis shows the bloom period in days. The y-axis is the number of pollen grains available for pollination. Consider that the number of pollen grains produced in a bloom period follows a bell curve, and so do the number of pollen grains picked up and carried by honeybees and the number of pollen grains needed for pollination.

Figure 4.

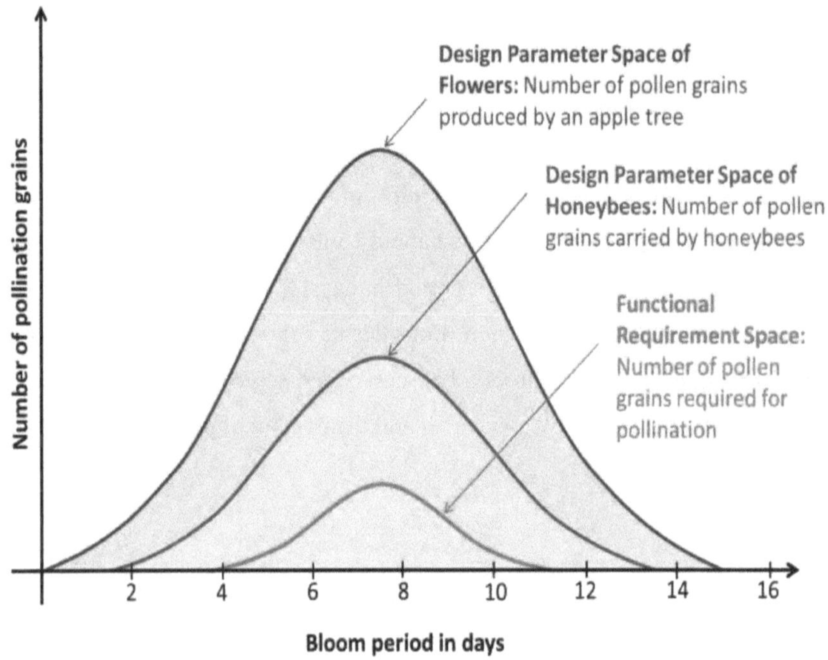

Functional requirement space, design parameter space of honeybees, and design parameter space of flowers, that enable the pollination process.

Let us apply the second axiom (i.e., the information axiom) of axiomatic design to the pollination process to check if it is robust. According to Way [125], a honeybee carries 100,000 pollen grains. Considering a factor of safety of 2, let us assume only half of them, i.e., 50,000 pollen grains on the bee‘s body, are available for pollinating a flower. Let us also conservatively assume that the probability of a pollen grain on the bee‘s body pollinating the flower is 0.001, i.e., 1 in 1000 opportunities. Consider each pollen on the bee‘s body as a Bernoulli trial, and the success outcome is the successful pollination of the flower. From the literature [125], it is known that an apple flower of many species needs only 5 to 20 pollen grains from another variety to set a full complement of seeds. If x is the number of grains that successfully pollinate the flower, the probability that more than 20 grains successfully pollinate the flower is given by

where $b(x;\ 50{,}000;\ 0.001)$ represents the binomial distribution with 50,000 trials and a 0.001 probability of a successful outcome in a trial. Recall that FR_0 = facilitate pollination process and DP_0 = flower–honeybee mutualism. The information content of the pollination process designed to pollinate successfully is given by

Since the information content of the pollination process design is close to 0, this design is simple, robust, and reliable; consequently, the pollination process has successfully produced seeds and fruits for millions of years. This longevity is enough to validate that the pollination process is robust.

Design Insights for Microrobots

A flower and a honeybee look like simple objects. However, they are complex and miniature structures with sophisticated anatomical and behavioral patterns that facilitate successful pollination processes. Honeybees, as a biological model, can offer a lot of clues for innovation. For example, bees use propolis, which is a sticky tree-resin substance, to seal their hive and protect the colony against pathogens. Despite its stickiness, honeybees can handle and manipulate propolis with their mandibles.

An experiment conducted by Saccardi et al. [126] indicated that the medial surface of the mandible is covered with a fluid substance that reduces propolis adhesion. This design clue can be used to solve the problem of resin deposition on woodworking machine tools. The axiomatic design approach presented in this paper offers an elegant methodology for designing efficient microrobots. Microrobots and nanorobots are incredibly small, requiring specific design features to function effectively. Miniaturization allows them to navigate small spaces, interact with microscale objects, or access confined environments. They can be used for targeted drug delivery, minimally invasive surgeries, or exploring inaccessible areas within the human body.

Insect-inspired microrobots mimic the locomotion and behavior of insects, such as bees, wasps, or ants. They are often used for tasks like environmental monitoring, search and rescue operations, or exploration of narrow spaces. Micro aerial vehicles (MAVs) are tiny flying robots that can navigate and perform tasks in the air. They find applications in surveillance, aerial photography, and environmental monitoring. There are nanoscale robots that are constructed using

DNA molecules as building blocks. They can perform specific tasks, such as molecular sensing or precise manipulation of biomolecules. Nano-swimmer robots are designed to move and navigate within liquids using various propulsion mechanisms. They can be used for tasks like cleaning contaminated fluids or nanoscale assembly. Kernbach et al. [127] described a meta-model of the re-embodiment of the biological aggregation behavior of honeybees in Jasmine microrobots in the context of the insect's sensor–actor system.

The flower and honeybee are based on cradle-to-cradle designs and are made of biomaterials [128]. At the end of life, they decompose and become biological nutrients to other organisms in their ecosystem. These biomaterials are upcycled, i.e., decomposed into reused components to create a product of higher quality or value than the original [129]. The design of microrobots, which may be millions and billions in number in a few years, must embrace two design strategies: (1) cradle-to-cradle design and (2) upcyclable design. If these design strategies are not followed, the world may suffer from microrobot and nanorobot junk and pollution. These microrobots dispersed in oceans, land, skies, and space may lead to serious environmental damage. Disassembling and recycling big robots used in industrial processes is easier than end-of-life management of microrobots. If a product passport for every microrobot is made compulsory by legal statutes, and 100% recovery recycling plans are in place, future microrobot pollution can be avoided. Adisorn et al. [130] described a digital product passport (DPP) implementation strategy as a policy instrument for a circular economy.

The design features of flowers and honeybees and their interactions can offer many insights into the design of microrobots and nanorobots. Nature uses 4D printing technology to create organisms such as honeybees and flowers. Ahmed et al. [131] discussed the fundamentals, materials, applications, and challenging elements of 4D printing. Aldawood [132] reviewed various 4D printing applications in electronics, renewable energy, aerospace, food, healthcare, and fashion. For example, a small single-camera imaging system that provides a continuous 280-degree field of view (FOV) inspired by the large FOV of insect eyes. These innovations are well-suited for mobile robots, particularly for flying vehicles that need lightweight sensors [133].

This camera is made with metal, glass, acrylic, and composites. However, these technical materials are challenging to recover, reuse, and recycle. There were many innovations based on insects. For example, Shi et al. [134] presented a hydrodynamic model of honeybee mouth parts that may inspire the design of high-frequency microjoints for engineering applications, such as microstirrers.

Table 2 presents some insights we could use to design microrobots, taking inspiration from honeybees, flowers, and the pollination process.

Table 2.

Insights from the design features of flowers and honeybees in the pollination process.

Design Strategy	**Flower**	**Honeybee**	**Insights**
Cradle-to-cradle	Yes	Yes	Design a microrobot with a cradle-to-cradle strategy.
Material choices	100% biomaterials	100% biomaterials	Most microrobots use materials such as metal, glass, rubber, plastics, and composites. It would be good if biomaterials were selected at the design stage.
Power source (Power autonomy)	Rely on food produced by photosynthesis using sunlight.	Rely on food produced by plants that are 100% environmentally benign.	Usually, microrobots use chemical batteries that are difficult and costly to produce and recycle. Microrobots can be designed to operate on bio batteries charged by solar and wind energy.
Miniaturization	All flower parts are small, measuring a few micrometers to a	All parts of a honeybee are small, measuring a few	Miniaturization technology is emerging, and a lot of research has to be conducted.

Design Strategy	Flower	Honeybee	Insights
	few centimeters.	micrometers to a few centimeters.	
Sensors	Biosensors	Biosensors	Biosensor technology is a well-researched topic, but a lot more to be advanced for their use in miniature engineered systems.
Actuators	Nature's forces, like wind and sunlight, help flowers to move (e.g., sunflower)	Muscles made primarily of carbon.	Artificial muscles made of biomaterials need to be developed.
Manufacturing techniques	Nanotechnology, 4D printing.	Nanotechnology, 4D printing.	More advancement in 4D printing is required.
Upcycle	All flowers are upcycled and recycled 100% by nature.	All honeybees are upcycled and recycled 100% by nature.	The design strategy should include recycling, upcycling, and reverse logistics with a product passport for every microrobot for recovery, disassembly, segregation, reuse, and recycling.

Learnings from the pollination process in nature can inspire design principles for microrobots aimed at mimicking the efficiency and adaptability of natural pollinators. Some potential design principles are as follows:

(1) Miniaturization: microrobots should be designed at a scale comparable to natural pollinators, allowing them to navigate through complex environments and access confined spaces within flowers.

(2) Agile mobility: incorporate agile and versatile movement mechanisms to navigate through intricate floral structures, similar to the flight agility of bees and other pollinators.

(3) Sensory systems: equip microrobots with advanced sensory systems, including cameras, proximity sensors, and environmental sensors, to detect and respond to changes in the surroundings, ensuring effective navigation and interaction with flowers.

(4) Bio-inspired adhesives: develop adhesion mechanisms inspired by the attachment strategies of natural pollinators, allowing microrobots to adhere to different surfaces within flowers without causing harm.

(5) Artificial intelligence: integrate artificial intelligence algorithms to enable autonomous decision making and learning from the microrobots, making them adapt to different floral structures and environmental conditions over time.

(6) Material compatibility: ensure that the materials used in microrobot construction are biocompatible with easy and natural recycling and upcycling.

(7) Environmental responsiveness: design microrobots to respond to environmental cues such as temperature, humidity, and light, similar to how natural pollinators adapt their behavior to changing conditions. By incorporating

these design principles, researchers and engineers can develop microrobots that mimic the efficiency and adaptability observed in natural pollination processes, contributing to advancements in fields such as precision agriculture, environmental monitoring, and ecosystem preservation.

Operations management is concerned with designing and controlling the production process and redesigning business operations to produce goods or services. Nature has also been operating using numerous and often efficient methods, such as pollination. The supply chain of a pollination process involves agents such as flowers and honeybees producing products such as nectar, pollen, honey, fruits, and seeds. The agents in nature‘s processes often play dual roles as customers and suppliers.

The requirement for FRs to be satisfied independently by DPs underpins the goal of creating designs that are simple, efficient, and capable of meeting user needs without unnecessary complexity or interdependence. By adhering to the independence and information axioms, designers can create systems that are effective, adaptable, and robust against changes and uncertainties.

Some DPs perform multiple tasks. A kitchen knife is used for peeling, cutting, chopping, dicing, slicing, mincing, and separating tasks. Similarly, the mandibles of a honeybee on the head sides act like a pair of pliers. The mandibles are used for any chores in a hive, such as grasping or cutting wax to construct the comb, biting into flower parts (anthers) to release pollen, carrying detritus out of the hive, or gripping/biting enemies during nest defense. This looks like a coupled design. However, a careful analysis will reveal that the DP fulfills multiple FRs simultaneously. Honeybees use mandibles (a single DP) to satisfy different FRs at different times, not simultaneously. Therefore, it is not a coupled design but an ideal design.

Without pollination, the genetic diversity of plant species would decline, leading to an unstable ecosystem. This paper presented an elementary analysis of the pollination process by using axiomatic design technology. The FRs of the pollination process that have been discovered in the last hundred years or so (and many more are yet to be discovered), when mapped onto the DPs that nature has already created, the pollination design seems to be an ―uncoupled design” which is the best design according to the theory of axiomatic design. However, the design equations presented in this paper are incomplete, and associations represent the high level of mapping of FRs and DPs. Much more research must be conducted to determine the specifications for all FRs and DPs. Only then can one conclude whether the pollination process is a good or suboptimal design. Because of the absence of specifications, the application of the second axiom (i.e., the information axiom) is rudimentary.

The axiomatic design methodology offers an elegant way of judging whether a design is good or suboptimal. In that sense, axiomatic design is a valuable tool. It can be applied to product design, service design, machine design, tooling design, software design, manufacturing system design, or logistics design. This paper focused only on the insights that can be derived from our observations and understanding of flowers, honeybees, and the pollination process to inspire microrobots and nanorobot designs. The field of microrobots and nanorobotics is still in its early research and development stage. However, they showcase potential applications and capabilities of miniaturized robots in various domains.

This was possible only at a high level because nature's pollination process is very complex with many interdependencies. This paper presents only an elementary analysis based on quantitative data presented by Way [125]. Within the limits of the data, this study is robust but incomplete. With advanced technologies such as the Internet of things (IoT) and video analytics researchers worldwide, the future might offer extensive data for studying the pollination process with axiomatic design.

Honey bee foraging behavior and their impact almond orchard

The European honey bee (*Apis mellifera,* thereafter honey bee) is an important insect globally for its ecosystem services and valuable products such as honey, bee wax, propolis, venom, and royal jelly. Together with other pollinating insects, honey bees offer pollination services for 85 to 90% of flowering plants to facilitate their reproduction. When bees land on flowers, their primary objective is typically to seek sustenance, specifically pollen grains and nectar, and unfortunately, it ends up in facilitating pollination activity. Pollen serves as an essential protein source for bees, while nectar provides the carbohydrates needed for energy [135-140].

Bees use these vital ingredients in the honey-making process. The diversity of plants that serve as forage resources for bees is vast and influenced by factors such as flower shapes, abundance, and the quality and quantity of nectar and pollen. These differences can impact honey bee foraging behaviour, colony health, the overall production of honey, and its quality. For example, nectar-rich flowers attract more foraging bees and often result in higher honey yields. Some plant species produce pollen that is richer in proteins, fats, and other nutrients essential for bee larval development than others. This intricate relationship underscores the importance of preserving diverse floral resources to support honey bee health, enhance their pollination efficiency, and optimise honey production [141-147].

Conservation of flowering plants is critical not only for sustainable honey production but also for supporting pollinators and maintaining ecosystem health. Because the quantity and quality of nectar and pollen tend to vary among plant species, it is crucial to identify and preserve key plant species, especially native flowering ones, for effective bee foraging and honey production in specific regions. Native plants, unlike exotic ones that may turn into invasive species, are well-adapted to local conditions and often provide superior food sources for native pollinators. In addition, a diverse range of flowering plants that bloom at different times throughout the year provides a continuous source of nectar and pollen for honey bees, preventing food shortages during critical periods. Therefore, fostering the growth and conservation of diverse, native flowering plants is fundamental to mitigating honey bee colony declines and ensuring the resilience of ecosystems that depend on pollination services [148- 154].

Conservation of flowering plants for pollinators and honey production in urban arid and semi-arid areas presents unique challenges due to the lack of enough rainfall. Land degradation in these areas has fuelled afforestation practices to restore ecosystem services. However, the majority of afforestation projects in arid and semi-arid areas normally select a few exotic tree species that grow fast and can withstand aridity, neglecting other important characteristics that influence the availability of nectar and pollen for pollinators and honey production. While drought-resistant exotic tree species are valuable for combating desertification, incorporating native flowering plant species into afforestation efforts ensures biodiversity and sustains pollinator habitats [155- 158].

Changes in floral availability caused by afforestation with exotic trees can disrupt plant-pollinator synchrony and reduce biodiversity. This can disrupt the synchronised relationships between plants and their pollinators. This can negatively impact the entire ecosystem, including native bees that rely on diverse plant species. Therefore, integrating native flowering species into afforestation efforts in arid and semi-arid areas is essential for fostering biodiversity, supporting pollinator populations, and ensuring sustainable honey production[151, 152].

Dodoma is a semi-arid region and one of the important honey production regions in Tanzania. Miombo wood, *Acacia,* and *Dichrostachys* shrubs dominate the natural vegetation of the region. This vegetation provides valuable nectar and pollen for honey production. However, the habitats are highly degraded due to the cutting of trees for charcoal production and building materials and overgrazing. In response, the government initiated an afforestation campaign to restore these degraded habitats. The University of Dodoma is one of the key stakeholders in the afforestation program, where more than 25,000 trees have been planted. Among the tree species planted, *Trichilia emetica* has been favoured for its drought tolerance and wide canopy, which provides shade [159- 161].

Although this program has contributed significantly to the conservation of animal species and carbon sequestration, replacing shrubs and herbs with a limited number of tree species could profoundly impact plant diversity, pollination services, and honey production. The aim of the study was to examine the role of plant diversity and abundance in supporting effective pollination and honey production by honey bees. The following hypotheses were proposed: (a) Foraging activities by bees are more noticeable in areas with a high abundance and diversity of flowering plants, and (b) native flowering plants support higher bee foraging activity than non-native species. These findings will provide valuable insights for the conservation and management of plant species critical to sustaining honey bee populations and enhancing their pollination services and honey production [161].

The Dodoma region has a semi-arid type of climate with an average annual rainfall of 400 to 570 mm per year and high geographical, seasonal, and annual variation [28]. Most of the precipitation falls from December to April, while there is a long dry season from May to November. The mean annual temperature ranges between 17 and 29 °C, with July being the coolest month and November being the warmest. The vegetation of the Dodoma region is composed of wooded grassland and bushes. The common plant species in the region are *Acacia tortilis*, *Dichrostachys cinerea, Adansonia digitata,* and Miombo species such as *Brachystegia microphylla* and *Brachystegia spiciformis* . The study site was located at the apiary centre of the University of Dodoma, situated at latitude 6° 12′ 53″ S and longitude 35° 49′ 29″ E. The natural vegetation of the university campus consists of short trees and shrubs. Exotic trees are planted in some areas for greening, shade provision, and other ecosystem services [162, 163].

Plant species identified in the sampling plots

The area's average plant density was 1,294 plants per hectare, predominantly shrubs and herbaceous plants, with a few trees. A total of 46 plant species were identified within the sampling plots. This plant diversity is typical of semi-arid regions, where vegetation must adapt to challenges such as low rainfall. Bees were seen foraging on a total of 22 flowering plant species in the sampling plots. About 75% of the foraging observations were recorded on non-native species, while 25% were recorded on native species. A statistical analysis revealed that this difference is statistically significant ($\chi^2 = 176.26$, d.f. $= 1, p < 0.001$). Additionally, approximately 58% of all foraging observations were recorded on shrubs, 30% on herbs, and 12% on trees. A statistical analysis revealed that the differences in foraging observations among shrubs, herbs, and trees were significant ($\chi^2 = 186.26$, d.f. $= 2, p < 0.001$). *Dichrostachys cinerea* was a dominant plant species in the apiary, where it formed continuous bushes (Fig. 5a). Open areas were dominated with herbaceous plants of the genus *Ipomoea* and *Cosmos sulphureus* (Fig. 5 b and c). Most of the plants that honey bees (*Apis mellifera*) gathered nectar from were shrubs, herbs, small trees, and crops grown in a nearby cleared space, such as passion fruit and sunflower. *Caesalpinia gilliesii, Acacia farnesiana, and Ipomoea pandurata* were the most foraged plant species. The majority of these plant species are non-native to the study areas.

Fig. 5

Some of the common vegetation in the study area: A. *Dichrostachys cinerea* shrubs B. *Ipomoea spp.*, C. A mixture of *Ipomoea* spp. and *Cosmos sulphureus* (orange flowers).

Factors influencing the number of flowers per plant

The rains began in December 2022, and the plant species flowered at different times of the year. The number of flowers per plant differed significantly between months ($\chi^2_{9, 348} = 66.153$, $p < 0.001$). Flowers were few in January and February but increased rapidly from March to June (Fig. 6). The number of flowers decreased from July, which is the beginning of the dry season. There was a slight increase in the number of flowers observed at the end of the dry season, coinciding with the sprouting and flowering of certain plant species during this period (Fig. 6).

Further, the number of flowers differed significantly between plant species ($\chi^2_{20, 348} = 76.248$, $p < 0.001$). Some of the plant species had flowers for only one month, while others flowered continuously for three months or more, including *Acacia farnesiana,* which had flowers for 5 months, and Dichrostachys *cinerea*, and *Caesalpinia gilliesii,* which had flowers for 4 months. Herbaceous species such as *Ipomoea pandurata, Ipomoea carnea,* and *Cosmos sulphureus* appeared towards the end of the rainy season, providing floral resources from May to July. Another factor that influenced the number of flowers was crown area, as trees with larger crowns produced more flowers. The number of flowers was significantly higher in plants with a big crown area ($\chi^2_{1, 348} = 40.579$, $p < 0.001$).

Fig. 6

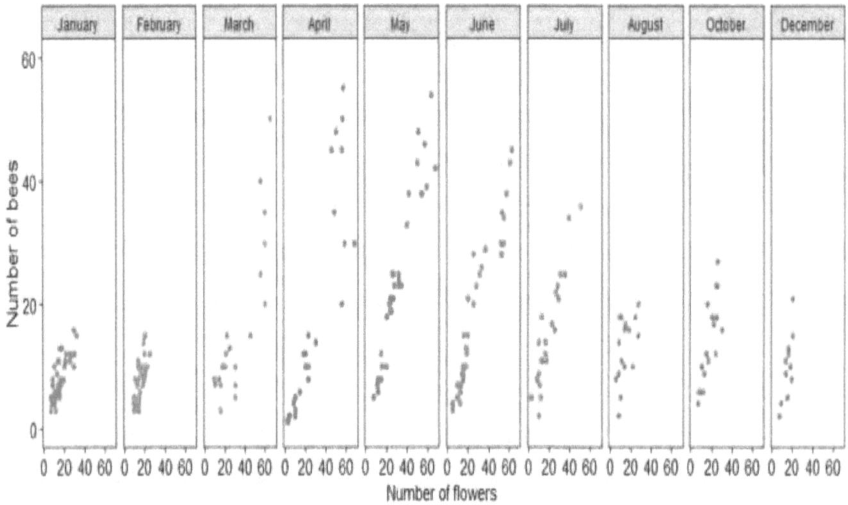

The relationship between the number of foraging bees observed and the number of flowers in different months

Factors associated with bee foraging

Bee foraging activities increased from March to May and started decreasing from June. A slight increase in bee activity was observed towards the end of the dry season in October, coinciding with the sprouting and flowering of some plant species during this period (Fig. 6). The difference in foraging activities between the months was statistically significant ($\chi^2_{9, 347=} 80.22, p < 0.001$). Bee activities also differed between plant species, where species that produce many flowers were often visited by a larger number of bees compared to species that had fewer flowers ($\chi^2_{21, 347=} 55.843, p < 0.001$) (Fig. 7). The number of foraging bees was positively associated with plant density ($\chi^2_{1, 347,} = 13.143, p = 0.004$) and plant diversity ($\chi^2_{1, 347,} = 11.847, p = 0.008$) in the sampling plots. Crown size of the plants was also significantly associated with foraging bee activity ($\chi^2_{1, 347,} = 13.942, p = 0.002$), with larger crowns attracting more bees (Fig. 8). Additionally, the number of foraging bees decreased with the distance of the sampling plot from the apiary ($\chi^2_{1, 347,} = 8.457, p = 0.013$).

Fig. 7

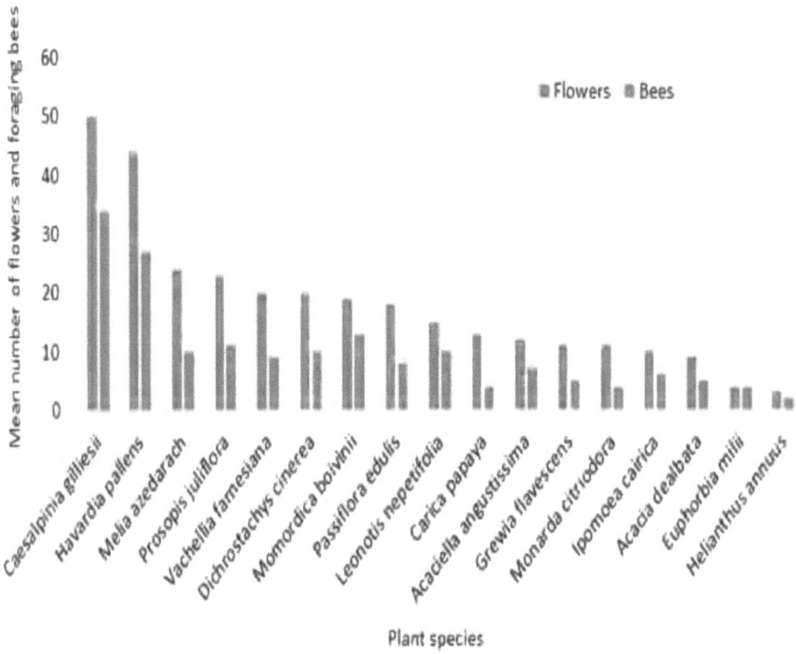

The mean number of flowers and foraging bees observed in different plant species in the study area

Fig. 8

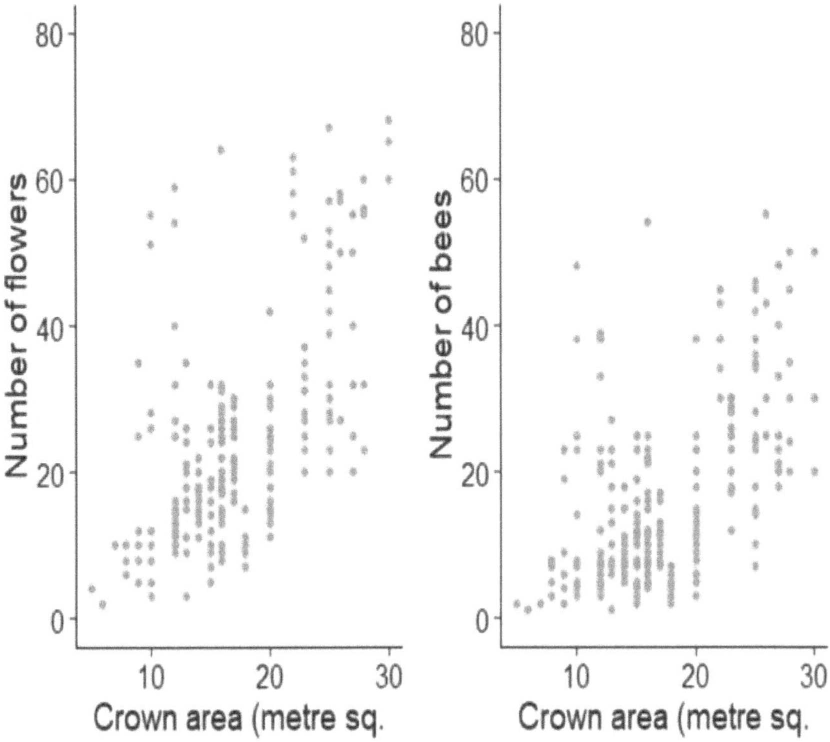

The association between crown area and the number of flowers and foraging honey bees. The number of both flowers and foraging bees increased with crown area

The findings from this study underscore the significance of diverse flowering plants in supporting honey bee foraging. While native plants have been shown to provide immediate foraging resources for honey bees, this study has shown the importance of maintaining a diverse range of flowering species regardless of nativity. A variety of flowering plants is crucial for ensuring a continuous supply of nectar and pollen, which directly impacts the health and productivity of honey bee colonies. The study highlights the need for conservation efforts that promote plant diversity, as this diversity is essential for sustaining honey bee populations and enhancing pollination services [154].

Caesalpinia gilliesii shrubs, though fewer in number, play a vital role in supporting honey bee foraging in the study area due to their prolific flowering. The finding aligns with a study by Borges et al. [170]. The plant is a perennial evergreen shrub native to subtropicl areas of Argentina and Uruguay but is now widely grown in warm regions around the world. It has exquisite flowers with yellow petals and freely exposed red stamens, which enhance bee foraging efficiency. *Dichrostachys cinerea* was the second plant species with a significant number of flowers and attracted a considerable number of foraging bees. The species is well-adapted to dry environments, flowering during critical periods when other food sources may be scarce. It produces bicoloured, fragrant flowers and is considered an important bee plant, especially for pollen collection. Its adaptation to aridity makes it an important plant species in the arid regions of central Tanzania [170- 173].

The findings indicate that as plant diversity increases, the number of bees also rises, highlighting the critical role that diverse floral resources play in supporting robust bee populations. This observation aligns with previous studies, such as those by Nicholls & Altieri [174], Isbell et al. [175], and Abbate et al.

[176], which demonstrated that a variety of flowering plants enhances foraging activity and promotes colony health, ultimately leading to higher honey yields. In contrast, monocultures, while they may offer abundant forage during peak bloom, often lead to resource scarcity once flowering ends. The lack of diversity can negatively impact bee populations, as they may struggle to find sufficient food sources later in the season.

High plant diversity prolongs the foraging season for honey bees. Some plant species bloom early in the rainy season, while others may flower during the mid to late rainy season or even at the end of the rainy season. With a wide range of plant species to forage from, honey bees can collect nectar and pollen over a longer period of time, enabling them to produce honey throughout the year. This different blooming time for diverse plant species ensures a continuous supply of nectar and pollen for pollinators. Promoting and protecting plant diversity is not only important for honey production but also for the overall health and sustainability of ecosystems [170- 180].

Overall, these findings underscore the importance of plant diversity in fostering healthy bee populations, which are vital for sustaining honey production and ensuring effective pollination services within agricultural ecosystems and honey production.

The bees were frequently seen foraging on wild plant species compared to the planted tree species. This finding agrees with Morandin and Kremen [181], Fukase and Simons [152], and Rahimi et al. [182]. Native plant species often provide a diverse range of floral resources, which increase the abundance and diversity of bees. Further native plant species are often better adapted to the local climate and can be more resilient in the face of climate change, providing a more stable and reliable source of nectar and pollen for honey bees and other pollinators [181-183].

However, the majority of foraging observations in this study were recorded on non-native plant species, indicating that the diversity of plant resources, whether native or non-native, is important for honey bee foraging. The finding aligns with those of Seitz et al. [151], Yang et al. [184], and [185], who emphasised the importance of non-native plant species for honey bee foraging, particularly in urban areas. The observation that honey bees predominantly foraged on non-native plant species underscores the importance of conserving a diverse community of plants, regardless of their native status[184, 185].

Nevertheless, the majority of afforestation programs in Dodoma use few exotic species of plants due to their faster growth compared to wild species. Although exotic tree species may provide valuable floral resources and other ecosystem services, the use of a few tree species in afforestation programs that bloom at the same time provides resources for bees only over a limited period of the year. A study by Liira and Jürjendal [186] found that the flowering period was longer in landscapes with a mixture of plant species compared to those with monocultures. Some bee species prefer to forage in open habitats with shrubs and herbaceous plant species. Furthermore, a high density of trees leads to less undergrowth, lowering the production of flowers by herbs and shrubs [188], which have been shown to be an important source of floral resources for honey bees in the study area. So, afforestation programs should consider the incorporation of a variety of tree species and the conservation of shrubs and herbaceous species [186-188] .

Bees are highly efficient foragers and are attracted to areas with a higher density of flowers [55]. When more flowers are available, bees can collect nectar and pollen more easily and efficiently [53, 56]. The association between flower abundance and bee abundance varied throughout the study period. During the peak of the flowering season (March to June), the number of foraging bees increased as they had access to ample food resources.

Conversely, during periods of floral scarcity at the beginning of the rainy season, in the dry season, or in areas with limited floral diversity, the number of foraging bees becomes low. This finding is supported by Hyjazie & Sargent [191] and Fowler et al. [192], who found a positive correlation between floral abundance and bee abundance. Thus, changes that reduce the number of flowers in an area are likely to negatively impact bee populations, as observed by Blaauw & Isaacs [153]. This aspect should be considered in afforestation programs to make sure that plant species that produce abundant flowers and high-quantity and quality pollen and nectar are identified and conserved [189- 192] .

Although honey bees can travel more than 10 km to search for desirable floral resources, most of the foraging activity is limited to about 5 km from their hive [193]. However, the bees were seen foraging frequently in areas within 0.5 km from the hives. It seems that the bees could find the resources at a shorter distance in this study, as the majority foraged within 3 km from the hives. This finding is consistent with previous studies by Couvillon et al. [194] and Hagler et al. [193], who noted that bees prefer to forage close to the hive, a behaviour linked to energy conservation. These patterns highlight the need to maintain diverse floral resources near hives, as this can enhance honey bee foraging efficiency and improve honey production[193].

Adoption of Sustainable Beekeeping Practices

Bees play a significant role in crop pollination. Bees are estimated to visit more than ninety percent of the world's top 107 crops. However, wind- and self-pollinated grasses account for about sixty percent of worldwide food production without animal pollination. Beekeeping is one of the sustainable agricultural activities that significantly contributes to rural development through its potential role in improving rural women's livelihood [195- 198].

Climate change, such as high temperature, is a big challenge for beekeeping sectors and food security. High temperatures could negatively affect insect species' populations through effects on survival, life cycle, fecundity, and dispersal. Variations in climate are known to influence the honeybee population, honeybee health, the pollination activity of insects, and their efficiency, with a major fall in bees and biodiversity. Moreover, global warming could also adversely affect plant–pollinator mutualism, and subsequently, temporal mismatches among the mutualistic partners might arise [199- **216**].

Beekeepers around the world have adopted several sustainable beekeeping management practices to adapt to climate change. Beekeepers improved their beehive boxes, transferred their beehives to suitable locations, and enriched the honeybees' food. Moreover, the adoption of different adaptation strategies such as changing the type and increasing the number of hives reduced climate effects [**218**]. Indeed, beekeeping management practices reduce climate stress. The selection of breeds and enhanced placement of hives according to climate situation, supplementary feeding during resource-scarce periods, and integrated pest management (IPM) to control insects, pests, and other stressors [[217- **219**].

Agricultural extension services play an important role in enhancing the knowledge needed to promote the adoption of sustainable beekeeping practices and foster the socio-economic status of rural women. Rural women can overcome barriers to adopting sustainable beekeeping practices by attending training and workshops delivered by the extension department. Saudi Arabia empowered the beekeeping industry through effective extension services and cooperative associations [**220- 222**].

The low dissemination of agricultural advisory services among women hinders women's decision-making and limits their involvement in agricultural entrepreneurial activities. Inclusive services are crucial for women's empowerment in agriculture. Agricultural extension services could empower women, promoting gender equality, rural development, and sustainable practices. They enhance farming expertise and cultural overcoming, disrupting traditional gender norms [**223, 224**].

Undoubtedly, gender equality and women empowerment are pivotal to gender transformation and a comprehensive approach to advancing gender in agriculture. According to Costa, the share of working women in agricultural employment increased from 56% to 64%, and men from 53% to 62%, across the countries. The legislatures of the democratic nations of Europe are fighting against discrimination and encouraging gender equality. Various stakeholders including public administrations, companies, and society are participating in this process. In 2021, women empowerment was a major concern for the 2030 Sustainable Development Goals (SDGs), especially at the United Nations Food Systems Summit [225- 227].

Moreover, the new Common Agricultural Policy (2023–2027) endorsed gender equality in rural areas. The different Member States and researchers discussed the issues faced by rural women and keenly focused on possible solutions to gender equality [**227,228**].

Agriculture contributes significantly to the economy of various countries around the world. In line with the Kingdom's Vision 2030, the management of available natural resources by adopting sustainable good practices, with emphasis on empowerment of rural women, is one of the major objectives of the KSA agricultural policy. The Kingdom of Saudi Arabia provides a great opportunity for women in different sectors to achieve economic security through a sustainable development strategy. Through collaboration between the Food and Agriculture Organization of the United Nations (FAO) and the Ministry of Water, Environment, and Agriculture (MEWA) of Saudi Arabia, the Sustainable Rural Agriculture Development (SRAD) program was designed and implemented. The program aims to develop capacities, enhance rural incomes of small-scale farmers, while conserving available natural resources. It covers various sectors, including coffee, rose, rainfed cereals, sub-tropical fruits, honeybees, fisheries, and livestock, involving various stakeholders in the countryside across 11 regions [**229**].

On 15 October 2024, FAO Saudi Arabia exhibited International Women's Day and demonstrated the impactful role of rural women in sustainable agricultural development. Moreover, Saudi REEF and the UN development program organized an event under the theme of Empowering for the Sustainable Future. This event provided income-generation opportunities for rural women through agricultural developments and small and medium enterprises. In January 2025, Saudi REEF released 2.4 billion in financial support to uplift rural communities across the Kingdom [**230, 231**].

Despite the harsh climate and arid conditions in Saudi Arabia, beekeeping is practiced on a large scale in many areas of the country. It is a significant source of income. The beekeeping industry is gradually increasing with several opportunities and challenges. Beekeeping not only improves families' livelihood but also grants a healthy sugar source, environmental sensitization, women empowerment, and social organization and plays a vital role in producing sub-products for healthcare and cosmetics. Implementing beekeeping programs required inclusive mechanisms for choosing suitable beneficiaries and appropriate ongoing extension, training, mentoring, and market access to improve the beekeeping industry [232- 235].

The Kingdom harvests more than 5000 tons of honey annually. The number of beehives surpasses one million denominations throughout the Kingdom. In addition, beekeeping practices are carried out on a large scale in Saudi Arabia. Currently, the country is dependent on imports. It has been estimated that more than 15,000 metric tons of table honey are imported from Australia, Turkey, Mexico, Argentina, Pakistan, USA, Germany, and Yemen to meet the demand. Generally, the prices of imported honey are lower than locally produced honey. However, Yemen supplies honey to Saudi Arabia, and local consumers are ready to pay up to USD 190 per kg. The Kingdom imports about 25,000 tons of honey annually. During the last three years, around 1.3 million parcels at SAR 130 million (USD 34.6 million) were delivered to Riyadh and used by beekeepers to produce honey [**236- 238**].

Since 2018, the Ministry of Environment, Water, and Agriculture (MEWA) launched various programs to promote the honey industry. Honey contributes 1.07 percent to the country's agricultural GDP. Six programs are running to support the beekeeping industry by developing infrastructure and local honeybee breeds. These programs focus on capacity building, regulating bee pastures, and stimulating investment and scientific research in beekeeping sectors [**236**].

In developing countries, rural populations are facing several challenges, including poor infrastructure, low agricultural productivity, vulnerability to natural disasters, climate change, increased aridity, poor market access, inconsistent government policies, poor gender integration, lack of proper agricultural extension service, unfavorable programs, and the cessation of initiatives developed by governments in developed countries as a result of the World Bank financial crisis [**239, 240**].

Undoubtedly, rural women play a significant role in developing countries, making up 45% of the agricultural workforce. Rural women in most parts of the world are involved in agricultural crop production and livestock breeding, providing their families with food, water, and fuel and contributing to non-agricultural activities to diversify their families' livelihoods. Many organizations and governments seek to empower rural women for the well-being of individuals, families, and rural communities. In the 1980s, the World Bank introduced women empowerment in the development process to fight against poverty and social injustice [**241-243**].

The Saudi government has launched an initiative to empower rural women by developing their technical skills in managing and growing small and medium enterprises to fulfill local market demand. The initiative focused mainly on capacity building in various areas of sustainable development, including farming and crafts, and providing important equipment to enhance the quality of small and medium enterprise in the rural areas across the Kingdom's regions. Attention to women's issues and their economic, political, and social empowerment has become a top priority of the Saudi leadership. The Kingdom's Vision of 2030 shows that the agricultural sector is a main pillar of sustainable development, where smallholders and producers, men and women, could play a vital role in rural agricultural development. The Ministry of Environment, Water, and Agriculture—facilitated through the Sustainable Rural Development Program, Food and Agriculture Organization of the United Nations—provides technical advisory assistance to develop the capabilities of rural population [244- 246].

The Saudi Reef program in KSA seeks to achieve environmental and agricultural sustainability by developing rural communities and providing job opportunities and livelihoods to small farmers. The Sustainable Agricultural Rural Development Program ―Saudi Reef" exclusively supports the bee sector by releasing SAR 140 million for 10,584 beneficiaries in various rural areas of the Kingdom. Saudi Reef‗s financial and technical support aims to promote local beekeeping to achieve self-sufficiency from honey. The Kingdom established royal bee breeding and stations in Hail, Najran, Jazan, Medina, Tabuk, and Taif to protect the bees from diseases and pests [**244**].

Beekeepers are provided with modern beekeeping tools and technologies; three mobile laboratories and four mobile clinics are working to examine and diagnose bee diseases and pests. In the kingdom region, 5850 licenses allow the people of 13 administrative regions, including 1560 beekeepers in the Hail. Therefore, the improved understanding of bee contributions to sustainable development is essential for confirming viable bee systems [**247**].

The respondents‗ socioeconomic and beekeeping characteristics. Most respondents (53.3%) were in the age group of 37 to 53. Only 40.2% of the respondents were less than 37 years old. Less than 5% of the respondents were more than 53 years old. Less than 5% of the respondents did not respond regarding their age. More than three-fifths of the respondents had obtained a university education. Only 32.6% of the respondents had basic education, and 4.3% were illiterate.

Over two-thirds of the respondents are involved in the beekeeping and processing industries. Specifically, 31.5% and 15.2% of the respondents practice beekeeping and processing, respectively. Less than 3% of the respondents did not respond regarding their occupation. Over three-fourths of the respondents were involved in full-time beekeeping and honey production. The remaining respondents were employed in the government sector (7.6%), private (2.2%) sector, or retired (5.4%). More than two-fourths of the respondents earned their primary source of income from beekeeping and being involved in trade and additional income.

Only 31.5% of the respondents were involved in beekeeping as a hobby. Only 1.1% of respondents did not respond to the purpose of beekeeping. More than four-fifths of the respondents were earning less than SAR 5000. Additionally, 14.2% of the respondents earned from SAR 5000 to less than SAR 15,000. However, 5.4% of the respondents did not respond regarding their monthly income. More than half of the respondents had two years or less than two years of experience in beekeeping.

Only 31.5% and 10.9% of respondents had up to six years and more than ten years of experience in beekeeping. Only 1.1% of the respondents did not respond regarding their experiences in beekeeping. More than 35% of respondents owned pastureland for beekeeping. In comparison, 25%, 26.1%, and 5.45% of respondents owned valleys, private farms, and protected land, respectively. More than three-fifths of respondents had a beekeeping practice certificate. Only 28.3% of respondents did not possess a beekeeping practice certificate.

More than three-fourths of respondents used modern hives. Only 15.2% of respondents used traditional hives. More than three-fourths of respondents had a hundred or less than a hundred hives. Only 16.3% of the respondents had more than a hundred hives. More than four-fifths of respondents produced 1000 kg or less of honey annually. Only 8.7% of respondents produced more than 1000 kg of honey annually.

All responses were arranged in descending order by the average score. The results indicated that the mean score ranged from a maximum of 3.99 to a minimum of 2.94. For the source of information regarding beekeeping, ―specialized agricultural websites" ranked first with an average score of 3.99. On the other hand, the ―universities" ranked last with an average score of 2.94.

All responses were arranged in descending order by the average score. The results indicated that the mean score ranged from a maximum of 4.15 to a minimum of 3.82. For professional guidance regarding beekeeping, ―productivity efficiency and honey quality" ranked first with an average score of 4.15. On the other hand, ―processing industries" ranked last with an average score of 3.82.

Beekeepers‘ needs for professional guidance and training regarding beekeeping. The majority of the respondents (71.74%) showed a high need for professional guidance and training regarding beekeeping, whereas 18.48% and 9.78% of respondents expressed a medium and low need for guidance and training, respectively.

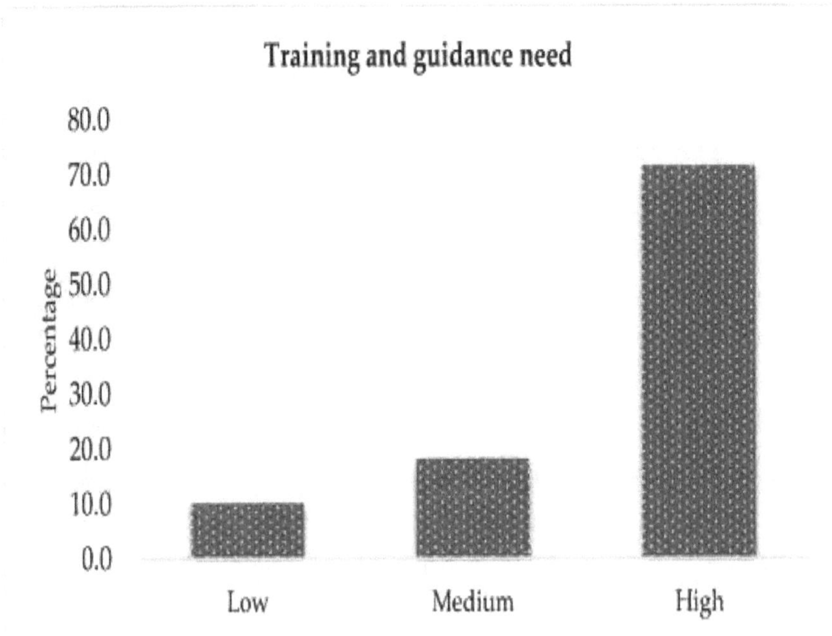

Figure 9. Beekeepers' need for professional beekeeping guidance and training.

The beekeepers' satisfaction regarding extension services. All responses were arranged in descending order by the average score. The results indicated that the mean score ranged from a maximum of 3.63 to a minimum of 3.30. For beekeepers' satisfaction regarding extension services, "awareness campaigns and agricultural extension caravans" ranked first with an average score of 3.63. On the other hand, beekeepers' satisfaction regarding the extension service "support for processed agricultural products manufacturing" ranked last with an average score of 3.30.

Figure 10 shows beekeepers' overall satisfaction regarding extension services. The majority of respondents (45.7%) showed a high satisfaction level regarding extension services, whereas 25% and 22.5% of respondents showed medium and low satisfaction, respectively.

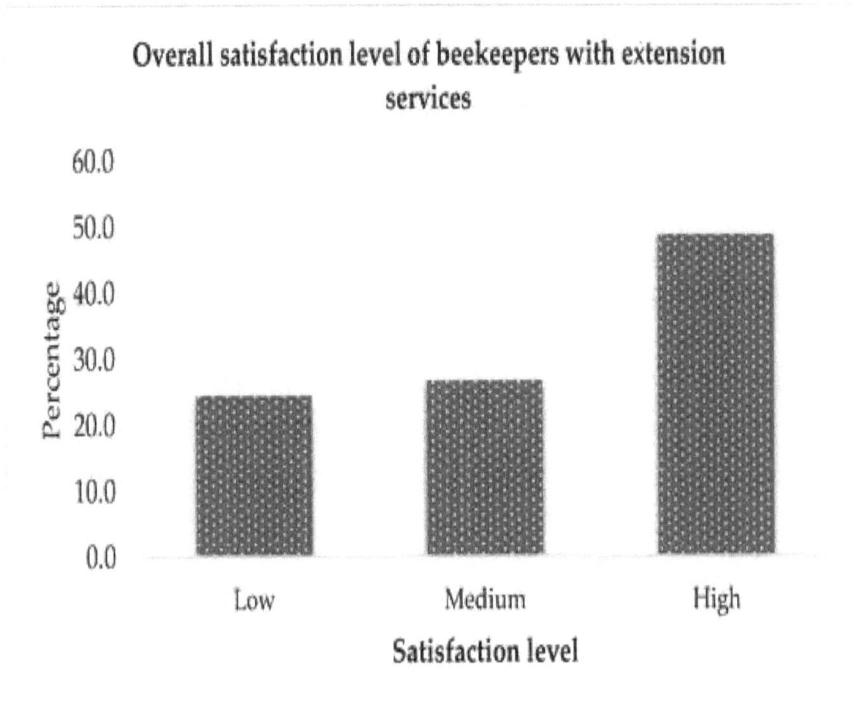

Figure 10. Overall satisfaction level of beekeepers with extension services.

The beekeepers‘ attitudes towards extension services. All responses were arranged in descending order by the average score. The results indicated that the mean score ranged from a maximum of 4.08 to a minimum of 3.66. For beekeepers‘ attitudes towards extension services, ―The Reef Program is beneficial and well-directed in supporting beekeepers" ranked first with an average score of 4.08. On the other hand, beekeepers‘ attitudes towards the ―Effectiveness in the institutions that adopt ideas and development, including the Rural Development Program" ranked last with an average score of 3.66.

Figure 11 shows the beekeepers‘ overall attitudes toward extension services regarding beekeeping. The majority of the respondents (43.5%) were neutral, whereas 22.1% and 31.4% of the respondents showed positive and negative attitudes towards extension services.

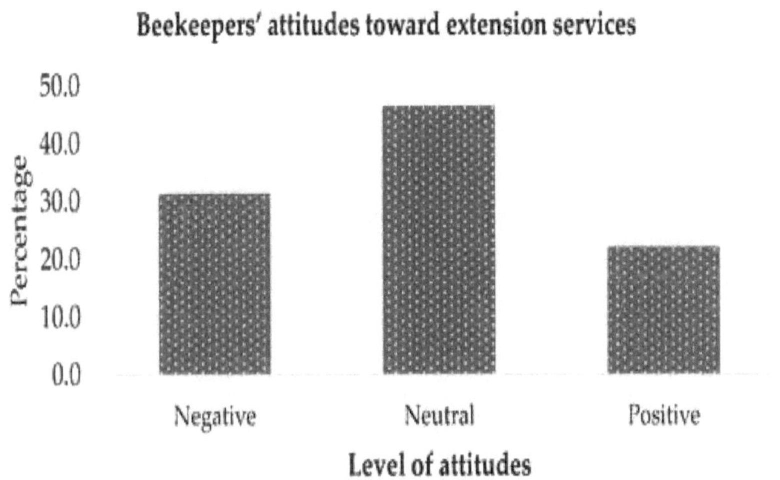

Figure 11. Beekeepers' overall attitudes toward extension services.

The industrial processing of beekeeping products. All responses were arranged in descending order by the average score. The results indicated that the mean score ranged from a maximum of 3.15 to a minimum of 2.43. The industrial processing of honey for –body care products” ranked first with an average score of 3.15. On the other hand, the industrial processing of honey for –royal jelly” ranked last by an average score of 2.43.

Figure 12 shows the overall industrial proficiency in manufacturing honey products. The majority of the respondents (51.1%) expressed high proficiency in honey products, whereas 31.5% and 8.7% of the respondents showed medium and high industrial proficiency in making honey products, respectively.

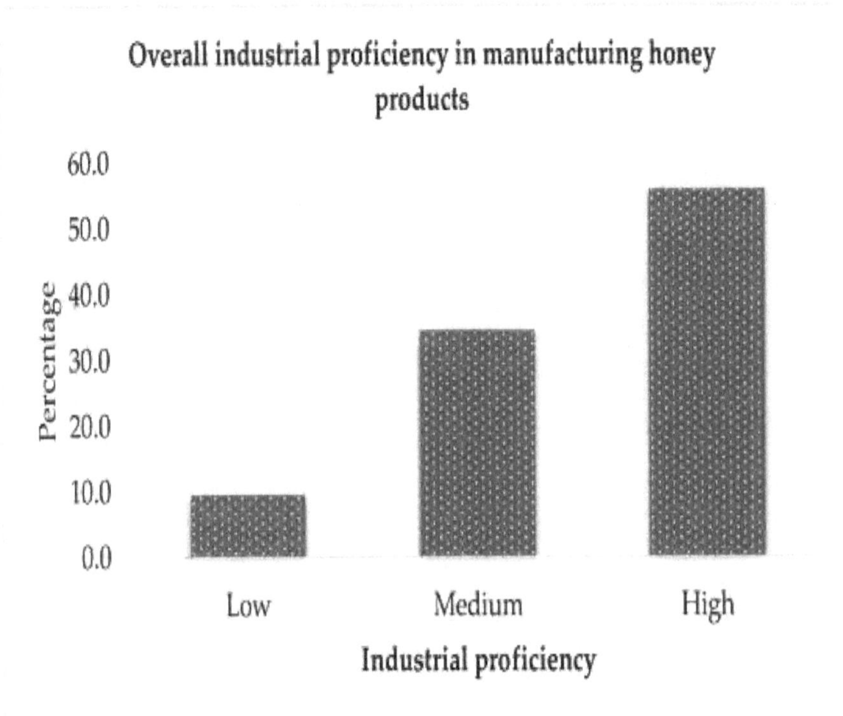

Figure 12. Overall industrial proficiency in manufacturing honey products.

The adoption of sustainable beekeeping practices among beekeepers. All responses were arranged in descending order by the average score. The results indicated that the mean score ranged from a maximum of 4.36 to a minimum of 3.75. For the adoption of sustainable beekeeping practices, –scientific methods are followed to protect bees from poisoning and pesticides" ranked first with an average score of 4.36. On the other hand, –Joinery of wooden beehives," ranked last with an average score of 3.75.

Figure 13 shows the overall adoption of sustainable beekeeping practices among beekeepers. The majority of the respondents (72.8%) adopted sustainable beekeeping practices at a higher level, whereas 19.6% and 2.2% of the respondents adopted sustainable beekeeping practices at a medium and low level.

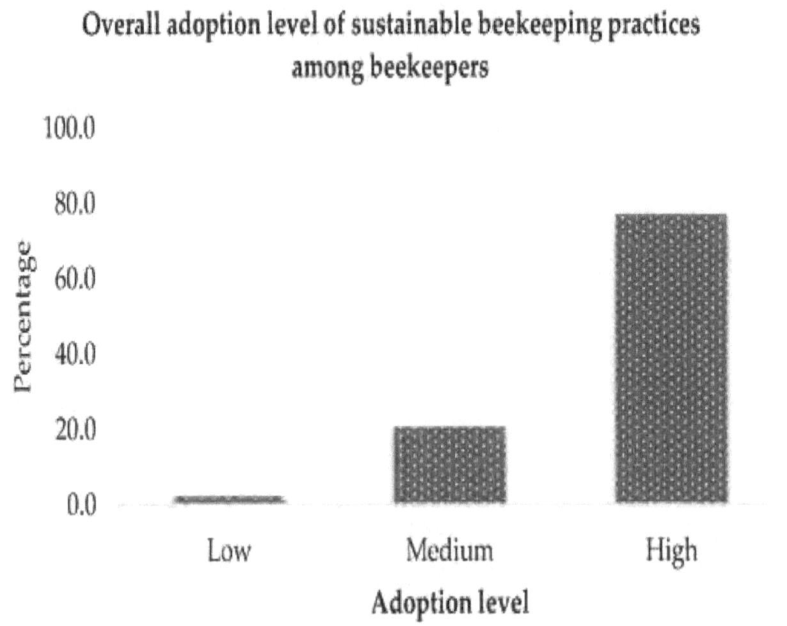

Figure 13. Overall adoption of sustainable beekeeping practices among beekeepers.

This research investigates beekeepers' satisfaction and attitudes toward extension services. In addition, it provides deep insight into sustainable beekeeping practices among beekeepers in Saudi Arabia. Findings revealed that most beekeepers were satisfied and presented neutral attitudes toward extension services. Those who participated in extension programs and training sessions gained entrepreneurial benefits.

Strong found extension agents disseminated innovative and sustainable beekeeping technologies to improve honey production in Kenya. Beekeepers who frequently visited the extension office were more aware of sustainable beekeeping technology. They had self-confidence about entrepreneurial behavior, and seemed more satisfied regarding extension services provided and economic gain. In the USA, Land Grant University provides online platforms like bee clubs, extension services, and apiary management technologies [**253, 254**].

Beekeepers in the study area might gain similar support, improving their satisfaction with extension services. Beekeepers might be highly dependent on extension agents for technical advice. Extension services would increase their satisfaction and build confidence in communication. Similar findings regarding the role of agricultural extension agents have been reported by various researchers. Our results contradict the findings of a qualitative study conducted by Benge and Amy, who revealed that commercial beekeepers were dissatisfied with extension services and being neglected when compared to hobby-type beekeeping practices. Commercial beekeepers have lost trust in extensions due to the beekeeping information that is published through the Florida Department of Agriculture and Consumer Services (FDACS), which is more for hobby beekeepers. Furthermore, they found that many extension agents showed less interest in commercial beekeeper education.[255-258]

The neutral attitudes of beekeepers toward extension services might be due to various factors such as poor extension services, lack of interest in extension services, and weak government policies. Our findings contradict with the findings of a qualitative study conducted by Gikunda et al. [**259**], which revealed negative attitudes toward honey beekeeping among women. The reasons behind the negative attitudes among women were cultural restrictions such as taboos, inheritance, and traditions. Moreover, youth learned honey beekeeping skills from their parents only. They demanded strong agricultural extension services regarding honey beekeeping technologies.

Anaeto et al. [**260**] reported that extension agents are commonly responsible for improving knowledge and skills and changing attitudes. The lack of access to external information negatively affects farm productivity and may keep beekeepers' attitudes neutral [**259-** 262].

Beekeepers in the study area might face similar challenges. In Saudi Arabia, there is a dire need to improve women's participation in beekeeping enterprises, extension training, and programs. Mburu et al. [**263**] found high participation of rural women in apiary management due to commercial insect programs, training, and the adoption of modern beekeeping practices.

Findings indicate high adoption of sustainable beekeeping practices among beekeepers. They might hold innovative information on the advantages of adopting sustainable beekeeping practices. It can be expected that beekeepers might have attended training sessions about advanced beekeeping techniques as the Bee Research Unit at King Khalid University and Aramco company initiated a training program entitled ―Beekeeping and Bee Products" and the Beekeepers Cooperative Society in Rijal Alma has a program which purposes to train 160 people [**264**].

The findings reveal that young beekeepers seem less likely to be satisfied with extension services. Low satisfaction with extension services among young beekeepers might result from poor extension services or low interest and reliance on traditional information instead of extension contact. Our findings contradict with that of Feketéné Ferenczi et al. [**265**], who conducted a qualitative study and found higher satisfaction regarding professional instructions about honey beekeeping. Young beekeepers reported that beekeeper workshops helped them in learning honey beekeeping practices.

Beekeepers' information varies according to conditions. Organized training, skills development programs, and management techniques should be provided to young beekeepers to improve their satisfaction. The extension department must investigate the reasons for low satisfaction with extension services provided to young beekeepers [**265, 266**].

Highly qualified beekeepers seemed satisfied with the extension services. Education among beekeepers is essential for successful beekeeping. Education builds capacity, belief, and improved attitudes. Educated beekeepers have a great ability to understand the objectives of extension services and the advantages of innovative beekeeping technology disseminated by extension agents for desired purposes [**267,268**].

A qualitative study conducted by Mulatu et al. [**269**] revealed that the educated beekeepers have more access to innovative information, and the possibility of attending professional training. They also influence farmers' adoption of contemporary hive beekeeping technologies and participate in income-generating diversification initiatives.

Beekeepers are facing various challenges due to irrelevant knowledge. A study conducted in the UK suggested that the government should make an effort to overcome these challenges. Meanwhile, the government of Saudi Arabia is striving to improve the technical knowledge of beekeepers through effective extension services [**270**].

Findings revealed that low dissemination of extension services caused negative attitudes toward extension services. A report prepared for the Sustainable Rural Agricultural Program (SARD) mentioned the advantages of the beekeeping sector in achieving the country's economic goals, as expressed in Vision 2030. Progress towards developing the beekeeper sector can be achieved through institutional support, and training extension staff at various levels [**238**]. Therefore, it is understood that poor extension service might reduce honey production and demotivate beekeepers towards extension services. The improvement in extension services among beekeepers was suggested in a report published by MEWA [**238**].

For instance, In sub-Saharan Africa, beekeepers improved beekeeping practices, adopted sustainable beekeeping practices through extension services, and initiated enterprises [**267**].

Moreover, it reported that beekeeping extension services were largely provided by NGOs and public agricultural extension departments. Public agricultural extension services and private consultants mostly delivered agricultural extension services for other agricultural enterprises to non-beekeepers. Extension service officers and/or NGOs critically promote the adoption and persistence of sustainable beekeeping. For instance, several extension services have been subcontracted by the Ugandan Government to private service agents to balance the operating costs of frontline agricultural extension delivery in remote areas [**271**]. This demand was made by extension service providers, who tend to favor training, skills development, and equipment provision to farmer groups rather than individual farmers [**271**-273].

The findings revealed that satisfaction with extension services among beekeepers improved their attitudes toward extension services. Understandably, the quality, availability, accessibility, diversity, relevance, and effectiveness of extension services ultimately built positive attitudes. According to researchers‘ observations, rural women started beekeeping for economic benefits. They stated that disseminating innovative beekeeping techniques could improve their knowledge and income. Rural women were completely dependent on extension services in terms of beekeeping. Only a satisfactory degree of extension services might improve their attitudes toward extension services. Verbeke et al. [**82**] reported that beekeepers‘ attitudes are driven through extension services and facilities to target markets by involving various stakeholders [274- 276].

The positive impact of honeybee activity

Sweet fennel (*Foeniculum vulgare*) is a plant belonging to the Apiaceae family. It is a short-lived perennial native to southern Europe and is often grown as an annual in cooler climates. There are two commercially important varieties of sweet fennel: *Foeniculum vulgare* var. dulce and *Foeniculum vulgare* var. vulgare. Sweet fennel is widely distributed in Africa and Asia, known for its potential as a natural and biological preservative in food science. Fennel is native to Southern Europe and the Mediterranean region and is cultivated throughout the temperate and subtropical regions of the world Fennel seeds are widely used for culinary purposes and have several uses in the pharmaceutical industry as well [277- 281].

Fennel contains various phytoconstituents, including flavonoids and phenolic compounds, which contribute to its medicinal properties . It is also used as a flavoring agent in food products and has traditional uses as an antispasmodic, diuretic, and antioxidant. It is commonly used in Egyptian cuisine and traditional medicine for its various health benefits [282- 284].

Sweet fennel attracts a diverse assemblage of beneficial insects, insects play a significant role as pollinators of fennel. They are essential for achieving good returns in fennel seed crop9. Various insect species, including honeybee (*A. mellifera*) bees, butterflies, and flies, visit fennel flowers for pollination , lacewings beetles bumble bees ants social wasps mud daubers, tachinid flies, and hoverflies, and the bee species, including *Apis florea*, *A. cerana*, *A. mellifera*, and *A. dorsata* were the most frequent visitors to fennel flowers13. The activity of different bee species on fennel varied with different abiotic factors, such as bright sunshine hours, temperature, and relative humidity. Which the fennel is a summer-blooming perennial that produces flowers attractive to a broad range of pollinators [285- 291].

Fennel nectar and pollen have an impact on bee broods. Honeybee (*A. mellifera*) collect and store pollen in colonies to ensure a strong worker bee population and brood production. The presence of brood stimulates pollen foraging in honeybee (*A. mellifera*), particularly in high pollen-hoarding strains. Pollen is an essential food source for bee larvae, providing them with protein, lipids, vitamins, and minerals. Phytochemicals found in nectar, honey, and pollen can affect bee infection and survival. Consumption of certain phytochemicals can increase antimicrobial peptide expression in honeybee (*A. mellifera*) and reduce levels of deformed wing virus (DWV)[292-295].

Insect pollinators have a significant impact on the quantity and quality of crop seed production. Different insect pollinators, such as honey bees and blowflies, have varying effectiveness in pollinating crops flowers, with honey bees being more effective. The presence of insect pollinators, such as honey bees, leads to higher seed yields and better germination rates compared to when inflorescences are caged or self-pollinated. The presence of insect pollinators during the flowering period of fennel leads to higher seed number and seed weight per umbel. In addition, seeds from flowers that were freely pollinated by insects showed higher germination capacity compared to those isolated from insect visitors. It also leading to higher seed set ratios and increased seed weights in plants such as sunflowers and rapeseed [296 – 299].

However, the specific impact of fennel nectar and pollen on bee brood was not mentioned in the previous works. This study aims to investigate the influence of fennel nectar and pollen on bee foraging broods, focusing on delineating the precise effects of fennel nectar and pollen on bee brood development, and the impact of *honeybee (A. mellifera)* on some fennel seed properties.

Diversity of insect visitors on fennel

Throughout the study period, a wide variety of insects were documented as they visited the blooming umbels of the fennel. The collected data revealed a total of six distinct species, spanning five different orders, seven genera, and six families. The observations unveiled diverse insect groups linked with fennel in the Qena region. Notably, these recorded pollinators include a range of bee species, such as honeybees, Carpenter bees, and parasitic wasps. Furthermore, the recorded observations reveal the presence of other insect groups contributing to fennel pollination. Hoverflies, beetles, seed bugs, and the pea blue were documented as frequent visitors, in association with fennel.

Impact of fennel on bees' activities

Foraging

Number of umbels visited/min

The presentation reveals variations in bee foraging behavior on fennel flowers, specifically in terms of the number of umbels visited per minute. The data indicates an increasing trend in bee activity over the three inspection weeks, with bees visiting an average of 36 ± 2.309 umbels per minute during the first week, 42 ± 2.667 umbels per minute during the second week and reaching a peak of 48 ± 3.464 umbels per minute during the third week (Fig. 14). This suggests an escalation in foraging efficiency and activity as the week progresses.

Figure 14

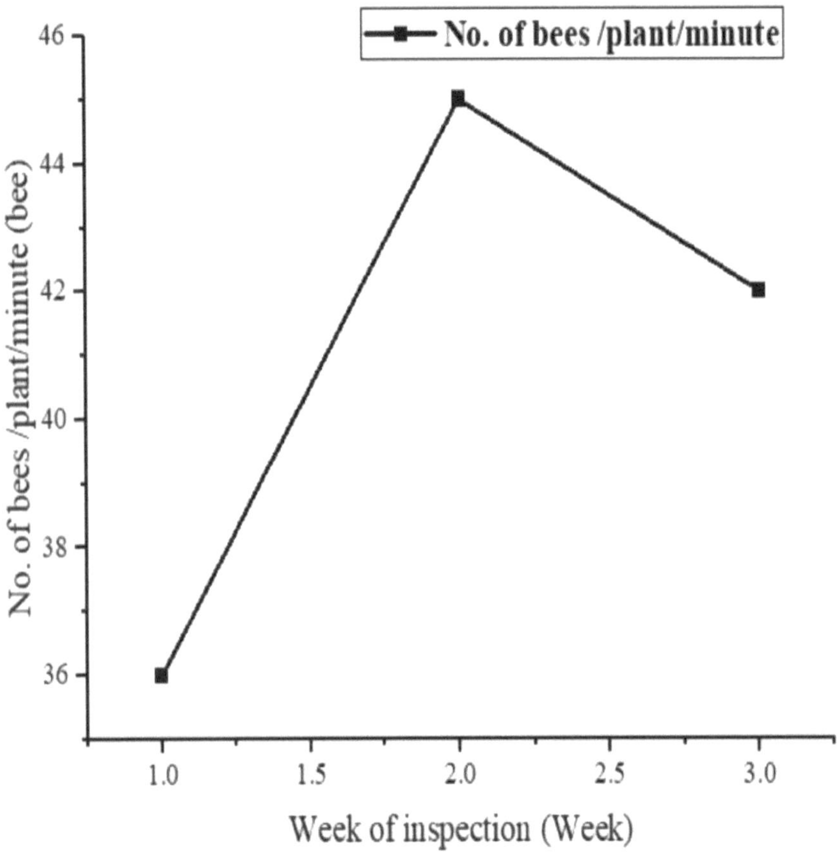

Relationship between fennel plant exposure and bee colony parameters.

One notable bee species, *Apis mellifera*, consistently demonstrates an elevated level of foraging activity throughout all inspection weeks. With a range of 36–48 umbels visited per minute, *Apis mellifera* exhibits remarkable efficiency in pollination, making it a significant contributor to the pollination process of fennel.

The statistical analysis supports these findings, with an R-squared value of 0.4778 indicating that 47.78% of the variation in umbels visited can be attributed to the different inspection weeks. However, the p-value of 0.1424 (at the 95% confidence level) suggests that the observed differences in bee activity across the inspection weeks are not statistically significant. The F value of 2.745 further supports this, indicating that the variations in umbels visited are due to random fluctuations rather than meaningful differences.

Fennel plant's impact on bee foraging and activity

Foraging outside the hive (Out): The data indicates that fennel plant exposure has a positive impact on the number of *honeybee (A. mellifera)* engaged in foraging outside the hive. Over the three-week period, the number of *honeybee (A. mellifera)* observed foraging outside the hive increased from 69.25 to 95.00 on average (Fig. 15). The R squared value (0.8811) indicates that approximately 88.11% of the variability in this foraging behavior can be attributed to fennel exposure. The extremely low p-value (<0.0001) indicates a strong statistical significance for this effect.

Figure 15

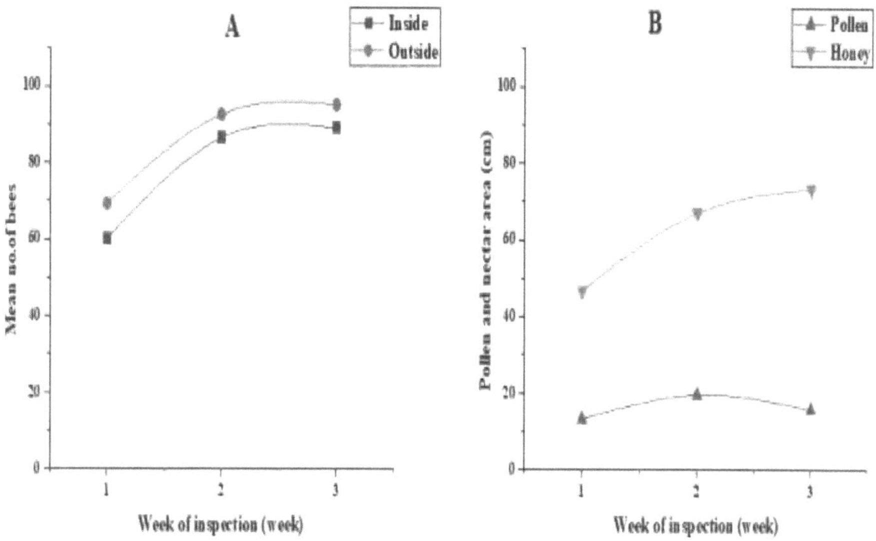

Relationship between fennel plant and bee foraging behavior (**A**), and food storage (**B**).

Activities inside the hive (In): Similarly, fennel exposure seems to positively influence the number of bees foraging inside the hive. The number of bees observed foraging inside increased from 60.00 to 89.00 over the study period (Fig. 20). The R squared value (0.8963) suggests that about 89.63% of the variability in this behavior can be explained by fennel exposure. The p-value (0.0001) highlights the statistical significance of this effect.

Pollen collection: The amount of pollen collected by the *honeybee (A. mellifera)* also appears to be influenced by fennel exposure. However, the R squared value (0.4958) suggests that fennel exposure explains about 49.58% of the variation in pollen collection, which is comparatively lower than the other parameters. The p-value (0.0459) is significant, but not as strongly as for other behaviors, indicating a moderate impact of fennel exposure on pollen collection.

Honey collection: Fennel exposure positively impacts the amount of honey collected by the bee colony. Honey collection increased from 46.75 to 73.25 over the three weeks. The R squared value (0.9016) suggests that approximately 90.16% of the variation in honey collection can be attributed to fennel exposure. The p-value (< 0.0001) signifies the strong statistical significance of this effect.

The impact of fennel plant colony activities

Sealed brood: The sealed brood refers to the capped cells containing developing bee larvae. The data demonstrates a significant positive impact of fennel exposure on sealed brood amounts. Over the three-week period, the sealed brood amount increased from 357.5 to 772.5 cells on average (Fig. 16). The high R squared value (0.9606) indicates that about 96% of the variability in sealed brood amounts can be attributed to fennel exposure. The extremely low p-value (< 0.0001) further reinforces the statistical significance of this result.

Figure 16

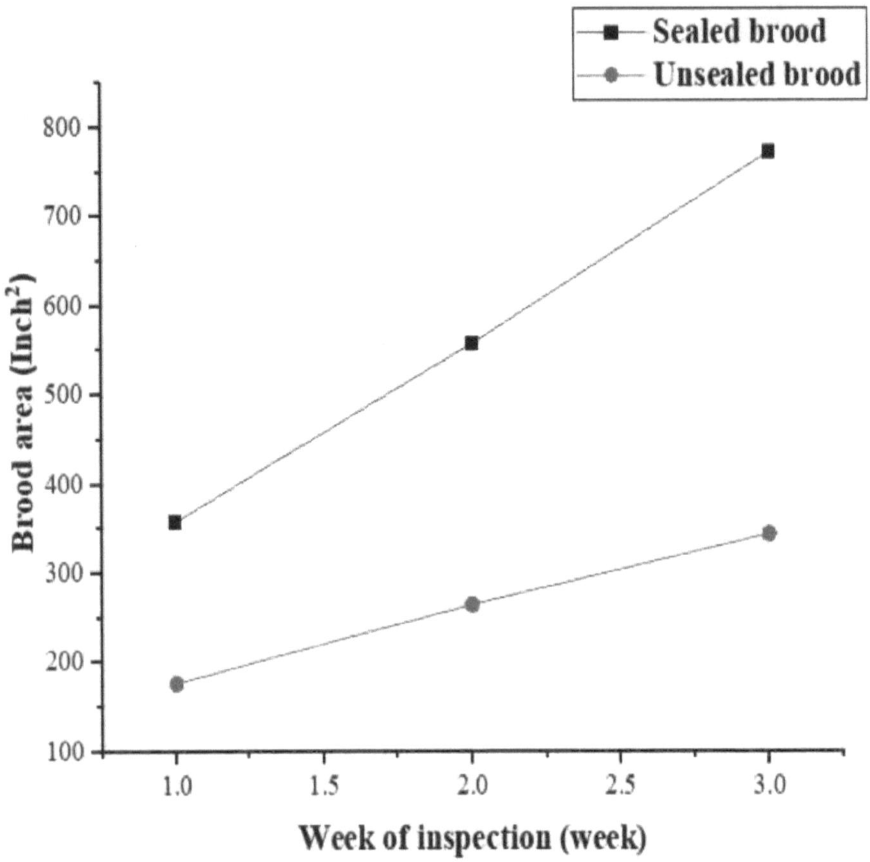

Effects of fennel plant on bee colony brood area (inch^2).

Unsealed brood: Unsealed brood includes the bee larvae in open cells. Similar to sealed broods, fennel exposure positively influenced unsealed brood amounts. The values increased from 176.3 to 343.8 over the three weeks. Although the R squared value (0.8478) is slightly lower than for the sealed brood, it still suggests a strong relationship. The p-value (0.0002) confirms the statistical significance of the effect.

Pollen: Pollen plays a crucial role in bee nutrition. Fennel exposure has a substantial positive impact on pollen collection. The pollen amounts increased from 53.25 to 257.5 over the study period. The R squared value (0.9874) indicates that fennel exposure explains about 98.7% of the variability in pollen amounts. The p-value (0.0459) is below the conventional significance level of 0.05 but is higher than for other parameters, suggesting a strong effect that may be influenced by other factors.

Honey: Honey is a vital energy source for bee colonies. The table shows a significant positive impact of fennel exposure on honey accumulation. Honey amounts increased from 257.5 to 877.5 over 3 weeks (Fig. 17). The high R squared value (0.9900) indicates that approximately 99% of the variation in honey amounts can be attributed to fennel exposure. The p-value (< 0.0001) underscores the robust statistical significance of this effect.

Figure 17

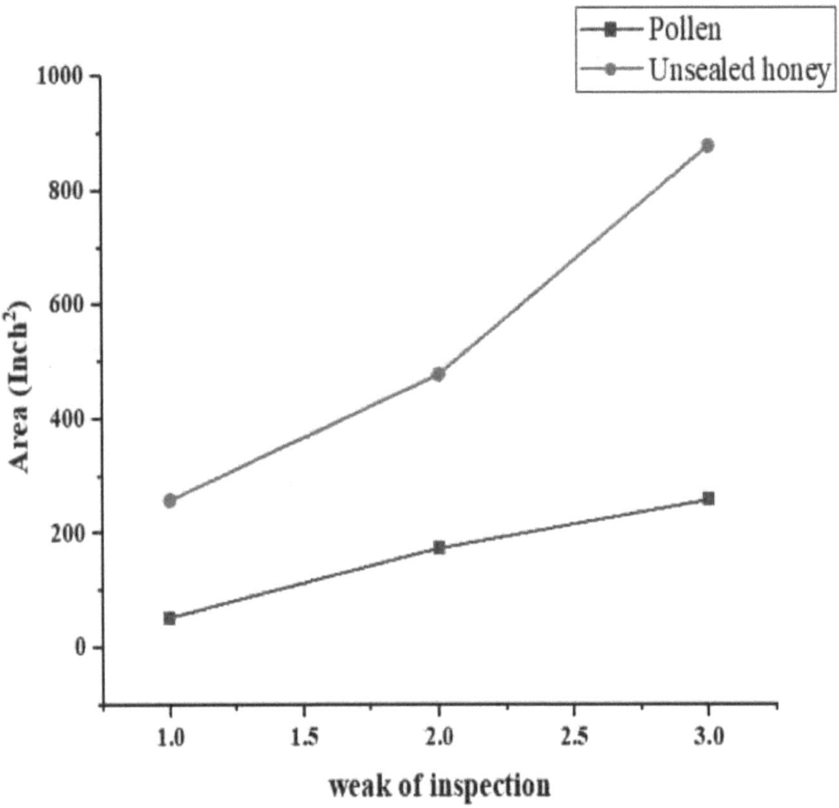

Increased honey accumulation in bee colonies with fennel: 3-week.

Colony strength: Colony strength is a measure of overall colony health and vigor. Fennel exposure is correlated with increased colony strength over the 3-week period. Colony strength values rose from 7.750 to 10 on average (Fig. 18). The R squared value (0.9313) indicates that about 93.1% of the variability in colony strength is explained by fennel exposure. The p-value (< 0.0001) reinforces the statistical significance of this effect.

Figure 18

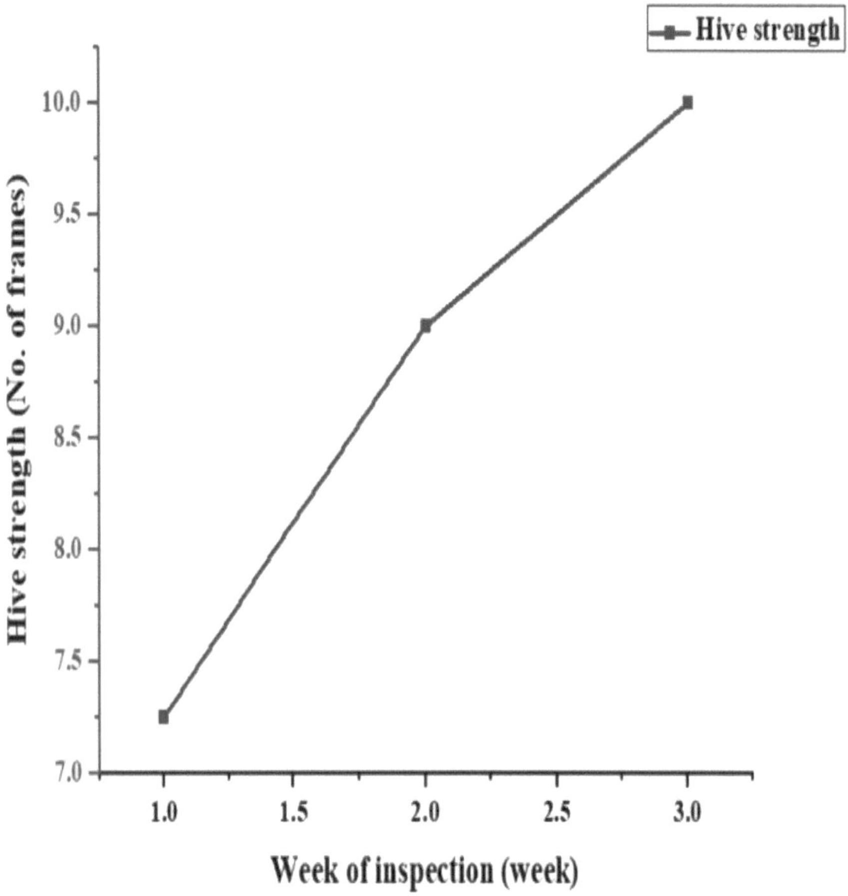

Enhanced colony strength with fennel exposure: impact over 3 weeks.

Impact of honeybee (*A. mellifera*) on fennel crop

Post-harvest quality, morphology, and commercial grades

Flower age and number estimation of seed on umbrella: Table 5 reveals the impact of bee pollination on fennel flower age and seed number of umbrellas. Bee-pollinated flowers have a higher average age (25.67) compared to self-pollinated ones (19.67), with significant differences indicated by the low p-value (0.0086). Additionally, bee-pollinated flowers have more seeds per umbrella (121.3) than self-pollinated ones (95.33), with a significant p-value (0.0163). These findings underline the positive influence of bee pollination on fennel reproductive characteristics and potential plant productivity (Fig. 19).

Figure 19

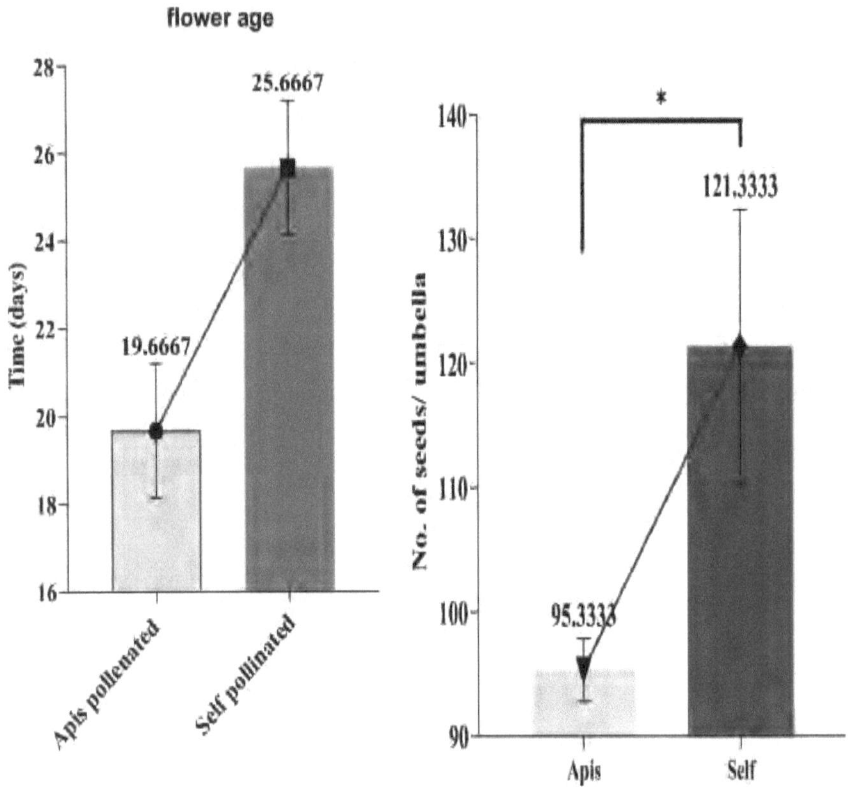

Comparison of fennel flower age and seed counts: apis vs. self-pollinated variants.

Seeds size and weight: The unpaired t-test data revealed a significant difference between open-pollinated and self-pollinated treatments in terms of physical parameters of fennel seeds, i.e., seeds size (length) and seeds weight. Where bee pollination has a significant positive impact on important plant seed characteristics compared to self-pollination. The mean difference in seed length is 2.133 mm (± 0.558 SEM), and for seed weight, it is 0.273 g/100 seeds (± 0.045 SEM) between the two pollination treatments.

The confidence intervals for both measurements indicate statistically significant differences between the treatments, as the confidence intervals do not overlap with zero, and the high R-squared values (0.343 for length and 0.906 for weight) suggest the pollination treatments explain most of the variation in seed characteristics. The p-values of 0.0007 (95% confidence) and 0.0034 (99% confidence) show highly significant differences between bee-pollination and self-pollination treatments. Bee-pollinated seeds tend to be longer and heavier.

Fruit weight per Umbel: In the realm of self-pollination treatment, the fennel fruit weight per umbel was recorded at 0.226 ± 0.0035 g (mean ± SD). Specifically, for the insect pollination treatments, the weight was measured at 0.619 ± 0.0324 g. In contrast, the treatments involving pollination without *honeybee (A. mellifera)* highlighted lower weights, with values of 0.226 ± 0.0035 g

The statistical analysis showed significant differences. It highlighted dissimilarities between both fennel pollination treatments (Chi-square = 7.13, df = 1, p-value = 0.008) as well as variations among the treatments (Chi-square 0.9733, df = 2, p-value < 0.0003).

Our rigorous statistical analysis revealed a compelling and significant difference between grain weight when *honeybee (A. mellifera)* were present and grain weight in the absence of *honeybee (A. mellifera)* ($p = 0.0034$, two-tailed). The grain weight means for the presence of *honeybee (A. mellifera)* (0.5100) exceeded the mean for the absence of *honeybee (A. mellifera)* (0.2367), with a remarkable difference of 0.2733 ± 0.04410 (95% confidence interval 0.3958–0.1509).

Seeds morphology

Seeds colors

The coloration of fennel seeds varied according to pollination treatment. Figure 21 visually illustrates the stark disparities between seeds produced from plants pollinated by bees and those from non-bee-pollinated plants. Bee-pollinated fennel seeds exhibited more naturally intense light greenish-brown pigmentation compared to self-pollinated fennel, which appeared in a grayish-green color, as well as a blackish-brown color. Thus, bee pollination Fennel yielded seeds are typically brighter, more intensely light greenish brown compared to self-pollinated fennel-yielded seeds, which were darker and greenish brown.

Seeds shape colors

The coloration of fennel seeds varied according to pollination treatment. Figure 20 visually illustrates the stark disparities between seeds produced from plants pollinated by bees and those from non-bee-pollinated plants. Bee-pollinated fennel seeds exhibited more naturally intense light greenish-brown pigmentation compared to self-pollinated fennel, which appeared in a grayish-green color, as well as a blackish-brown color. Thus, bee pollination Fennel yielded seeds are typically brighter, more intensely light greenish brown compared to self-pollinated fennel-yielded seeds, which were darker and greenish brown.

Figure 20

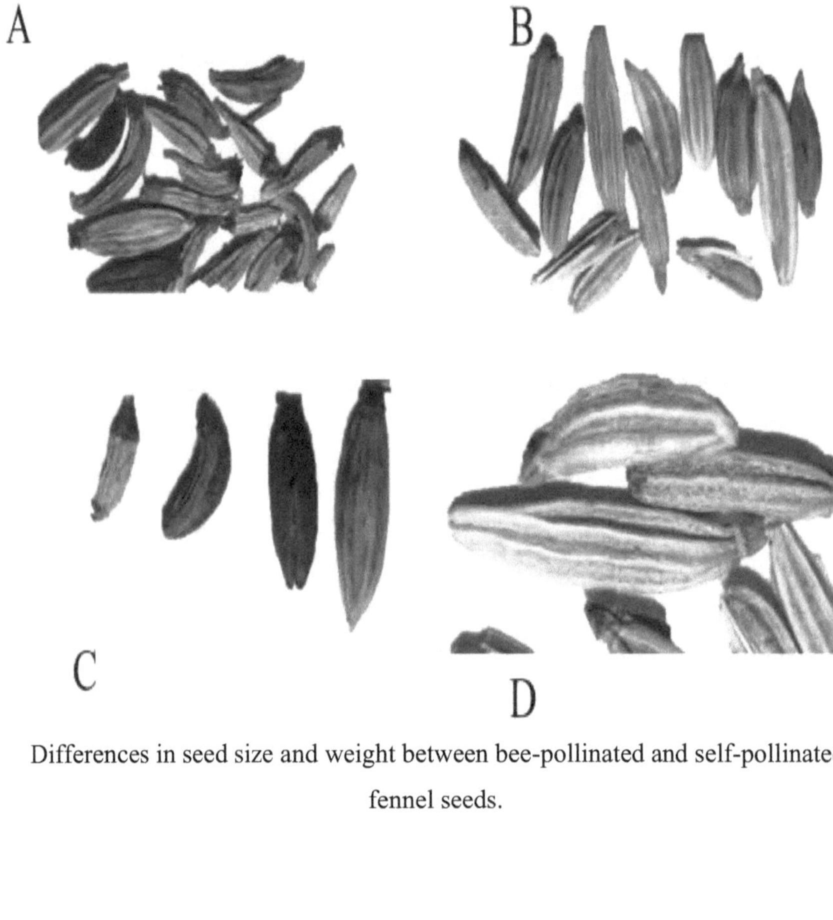

Differences in seed size and weight between bee-pollinated and self-pollinated fennel seeds.

Regarding shape, Seeds derived from bee pollination exhibited remarkable shape uniformity, demonstrating the consistent influence of this pollination method. In contrast, significant disparities were evident in the shapes of grains resulting from self-pollination in plants, indicating a distinct and notable effect of the pollination process (Fig. 21).

Figure 21

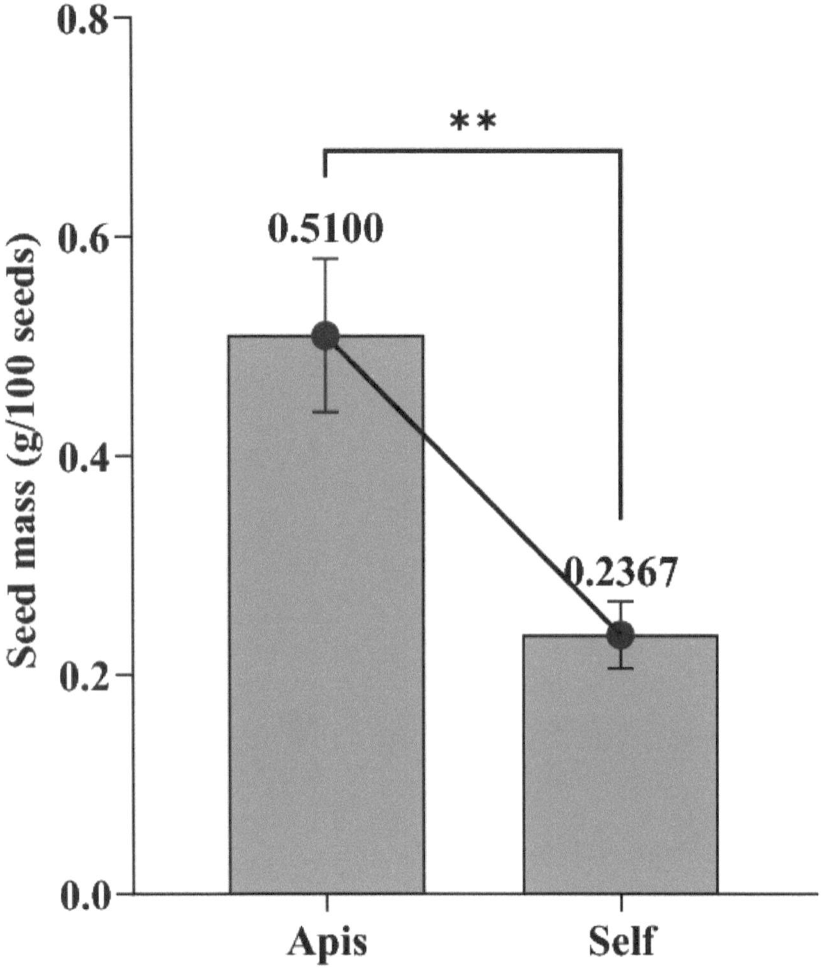

Differences in seed size and weight between bee-pollinated and self-pollinated fennel seeds.

Insect diversity and attraction by fennel

During the duration of our investigation, we documented a wide range of insect visitors to the blooming umbels of the fennel plant. This comprehensive examination revealed a diverse array of pollinators, including multiple bee species such as honeybees (Apis mellifera), carpenter bees, and parasitic wasps, as well as hoverflies and ladybug beetles. These findings align with research conducted by Layek et al.[302], who reported a high diversity of floral visitors on fennel crops in India, highlighting the importance of Apis mellifera as a key pollinator. Similarly, Kumar and Singh elucidated the significant influence of insects on fennel crop pollination dynamics, emphasizing the pivotal role of honeybees.

In Hisar, India, Shilpa et al.[286] identified 25 insect species from five orders visiting fennel flowers, with bees such as *Apis florea*, *A. cerana indica*, *A. mellifera*, and *A. dorsata* being the most frequent visitors. Furthermore, Bharti et al.[285] identified 25 insect species from 15 families and five orders as pollinators of fennel flowers, noting the high efficiency of *A. mellifera* as a pollinator [302, 303].

Fennel flowers possess traits that are highly attractive to bees. These traits include abundant nectar and pollen rewards, bright clusters, and strong scents that guide bees efficiently. The shape of fennel flowers ensures effective pollen transfer during feeding, maximizing cross-pollination. Fennel blooms coincide with periods of high bee activity, providing ample resources when competition is low, making fennel an optimal choice for bees. These characteristics enhance resource availability and reproductive success for both fennel and its pollinator bees.

Impact of fennel on bee activities

Bee foraging behavior is influenced by numerous factors, including environmental conditions, floral resources, and foraging strategies. Our study reveals variations in bee foraging on fennel flowers, particularly in umbel visitation frequency. We observed an increasing trend in bee activity over three weeks, indicating heightened foraging efficiency.

Apis mellifera consistently demonstrated significant pollination prowess, making it a crucial contributor to the pollination process of fennel. The statistical analysis supports these findings, showing that fennel exposure significantly impacts bee foraging outside the hive, accounting for approximately 88.11% of the variability observed. Inside the hive, fennel exposure positively influences pollen collection and honey accumulation, with significant increases documented over the study period.

Several studies corroborate the impact of fennel on honeybee foraging behavior. For example, Carisio et al.[304] emphasized the substantial influence of insect foraging on nectar productivity in flowers, suggesting shorter-term nectar volume measurements as a reliable assessment method. Layek et al.[302] highlighted the high number of pollen grains carried by honeybees on their body surfaces, reinforcing the significant role of fennel in enhancing pollination efficiency .

Impact of honeybee pollination on fennel

Our study underscores the pivotal role of honeybee (*Apis mellifera*) pollination in enhancing fennel (*Foeniculum vulgare*) crop productivity. By meticulously monitoring various parameters, we observed significant improvements in both the qualitative and quantitative aspects of fennel plants exposed to honeybee activity compared to those excluded from pollination.

Our primary observations indicated a substantial increase in seed set, seed weight, and overall crop yield in bee-pollinated fennel plants. Specifically, the seed set in these plants exhibited an approximate increase of 25%, and the seed weight showed significant enhancement. This improvement is attributed to the high pollen transfer efficiency facilitated by honeybees, which spend an average of 3.2 s per flower, ensuring effective pollination and fertilization rates.

In contrast, fennel plants isolated from honeybee pollination exhibited lower seed set and reduced seed weight, highlighting the crucial role of honeybees in enhancing crop productivity. Statistical analysis confirmed these findings, demonstrating a strong positive correlation ($r = 0.85$, $p < 0.01$) between honeybee visitation frequency and fennel seed yield.

Our findings align with and expand upon existing literature. Layek et al.[302] reported that honeybees are primary pollinators for fennel crops in India, significantly contributing to improved crop yield. Their study noted a 20% increase in seed set due to honeybee pollination, closely mirroring our observed increase. Similarly, Kumar and Singh[303] emphasized the critical role of honeybees in enhancing fennel seed quality and quantity, attributing these benefits to efficient pollination services provided by honeybees.

Shilpa et al.[286] documented that fennel plants visited by honeybees showed a 22% increase in seed yield compared to those not visited by bees, aligning with our findings, and highlighting the consistency of honeybee pollination benefits across different geographic locations. Additionally, Bharti et al.[285] identified *Apis mellifera* as the most effective pollinator among various insect species, supporting our observations regarding the superior pollination efficiency of honeybees.

Ecological and agricultural implications

The ecological interaction between honeybees and fennel plants not only benefits crop productivity but also enhances environmental sustainability. Honeybees facilitate cross-pollination, contributing to genetic diversity and resilience in fennel populations. This mutualistic relationship ensures that fennel plants receive necessary pollination services to produce viable seeds, while honeybees benefit from the abundant nectar and pollen resources provided by fennel flowers.

The study reinforces the importance of conserving honeybee populations and promoting practices that support pollinator health. Sustainable agricultural practices incorporating bee-friendly habitats can lead to increased crop yields and improved agricultural sustainability. By fostering an environment conducive to honeybee activity, farmers can enhance the productivity of fennel and other crops, ensuring long-term agricultural success.

Relevance in Conservation and Sustainable Agro-Ecosystem

Plant and pollinator interaction results in the pollination of various plants that are *self-incompatible*. These pollinating agents are important for the existence and reproduction of about 87.5 percent of wild species of plants. Pollinators are important to increase agriculture production and in ecosystem functions to increase plant genetic diversity. The total annual economic value of crop pollination around the world has been estimated at about €153 billion [306-310].

Klein et al. [309], surveyed that more than 87 of the world's leading food crops, representing 35 percent of global food production, depends upon animal pollinators, pollination by insect directly contributed about $20 billion and that of honey bee contribution was $14.6 billion in 2000 in the U.S. The total economic value of insect pollination of Chinese fruits and vegetables recorded about52.2 billion US dollars in 2008, which represented 25.5% of the total production value of the 44 crops produced in China.

The area among the pollinator-dependent crops has increased up to 300% during the last 50 years, both in temperate and tropical crops. Pollination is an essential prerequisite to seed and fruit development; it is a pivotal, keystone process and insects provide an important function in both natural and managed ecosystems. The mutual relationship has been found between insect pollinators and flowering plants, that in return nectar and pollen are major food rewards for pollinators [311- 316].

Robbins et al. [317] reported that most accepted estimates indicate honeybees' account for at least 80% of all insect pollination. For decades the consequences of insect pollination have been documented in treaties, McGregor, and Pesson and Louveaux [320]. In apiculture the most important species, *Apis mellifera* L. has been reported as the single most important pollinator. The pollinating potential of a single honeybee colony becomes evident when it is recognized that its bees make up to 4 million trips per year and that during each trip an average of 100 flowers are visited [317-321].

Pollinators diversity in agro-ecosystem

An enormous number of the world's insect diversity visits flowers for nutrition, but all are not efficient pollinators. Among crop foods, fibers, edible oils, medicines, and other valuable products, a significant production occurs due to the vital role of insects and other animal pollinators. In all types of ecosystems, bees are recognized as the most valuable pollinators, but their precise roles in pollination are not well documented. According to estimation, approximately 30,000 species of bees are known in entomology, and about 190 species of bees have been reported to be associated with pollination in North America. Some of the other noteworthy contributors in pollination are; alkali bee (*Nomia melanderi* Cockerell) found as solitary bees to pollinator in alfalfa and alfalfa leaf cutting bee (*Megachile rotundata* Fabr.) also play its role in the pollination of this crop. Through artificial mud nesting tunnels, orchard bees (*Osmia* spp.) can be managed in the field for pollination, as orchard bees are far much better for apple pollination than honeybees [322- 326].

Another, bumblebees (*Bombus* spp.) are pollinators of red clover (*Trifolium pratense* L.) and cranberries (*Vaccinium macrocarpon* Ait.), but difficult to manage in field conditions. Carpenter bees (*Xylocopa* spp.) are the outstanding pollinators of vine crops, especially of passion fruit (*Passifloraedulis* Sims), giant granadilla (*P. quadrangularis* L.), kiwi-fruit (*Actinidia deliciosa* (A.Chev.) C.F. Liang & A.E. Ferguson), various gourds, and winged beans (*Psophocarpus tetragonolobus* (L.)) but there is not much available data on these pollinators [327].

The interdependency of plants and pollinators vary in their degree, some plants species depend primarily on a single species of pollinator, which in turn has restricted sources of pollen or nectar. One example of a closely dependent association is the interaction between plant Yucca (Agavaceae) and its pollinators, yucca moth (*Tegeticula* spp.), having a mutualism estimated to be more than 40 million years old. Squash crops i.e. pumpkins and gourds are pollinated by specialized squash bees, *Peponapis* spp. and *Xenoglossa* spp., and are more manageable, they nest in underground burrows and become active at dawn, visiting cucurbit flowers until about midday when unisexual flowers close [328- 330].

There are many other pollinator native bees, which include sunflower bees (*Eumegachile pugnata* (Say), blueberry bees (*Habropoda laboriosa* (Fabr.)), and *Osmia ribifllloris* Michener (Table 4). Oil palm weevil, *Elaeidobius kamerunicus* as the most valuable in terms of the economic importance of this crop at the world trade level. Non-biting midges, *Forcipomyia* spp., a specialty pollinator among the cocoa (*Theobromacacao* L.) crop, which breeds in rotting vegetation in its plantation are a few of specific examples of pollination to produce the economic plants/crops that are in need of human beings [331- 337].

Table 4.

Bees and specific plant fauna they visit and pollinate.

Common name	**Scientific name**	**Example of crop plants pollinated**
Alkali bee	*Nomia melanderi*	Alfalfa, clover, mint
Blueberry bee	*Habropoda laboriosa*	Blueberry
Carpenter bee	*Xylocopa* spp.	Passion flower, eggplant, pepper
Digger bee	*Andrena*, *Colletes*, and *Melissods* spp.	Cotton, fruit trees
Alfalfa leafcutting bee	*Megachile rotundata*	Alfalfa
Blue orchard bee (mason bee)	*Osmia lignaria*	Almond, apple, sweet cherry
Squash and gourd bee	*Peponapis pruninosa* other *Peponapis* and *X enoglossa* spp.	Squash, pumpkin, gourds
Sunflower bee	*Eumegachile pugnata*	Sunflower

Among the herbivorous insects, the interaction of butterflies and moths is found during both its larval and adult stages and the latter is involved in pollination (Table 5). Some of these are *Heliconius* butterfly; yucca moth, obligatory mutualisms are exceptional in order Lepidoptera of pollinators. Moths from the families Geometridae, Noctuidae, and Sphingidae are among the most studied moths to be known as pollinators, their pollinating activity takes place at night in many plants such as cacti, orchids, trees. Many had also identified and reported thrips on flowers and they noticed that thrips feed on pollen so that they can be effective pollinators or minor or secondary pollinators of a wide variety of plants in agro-ecosystem or nature [338- 357].

Table 5.

Beetles, butterflies, and moths with specific plant fauna they visit and pollinate.

Plant	Pollinator insect	References
	Beetles	
Asimina	*Euphoria sepulchralis*	[338, 339]
	Trichius spp.	
	Trichotinius lunulatus	
Calycanthus	*Calopterus truncates*	[340]
Calochortus	*Acanthoscelides* spp.	[341]
Linanthus	*Trichochrous* sp.	[342]
	Butterflies	
Wild carnation, *Dianthus carthusianorum*	Butterfly species	[343]
Native plants of North America	Checkerspot butterfly, *Euphydryas editha bayensis*	[344]
Milkweed and other	Monarch butterfly, *Danaus plexippus*	[345]

Pollinators and plant interaction

In an ecosystem, the interaction between the organisms favors co-evolution and it gradually helps to evolve together for some betterment or for existence in nature. Those plants having a generalist pollination system, have a link between pollinator diversity and seed production can also be established. Pollination biology (Figure 22) draws attention to both evolutionary and ecological approaches i.e. the link between pollinator behavior and plant mating patterns, generalization, and specialization in a pollination system [358- 360].

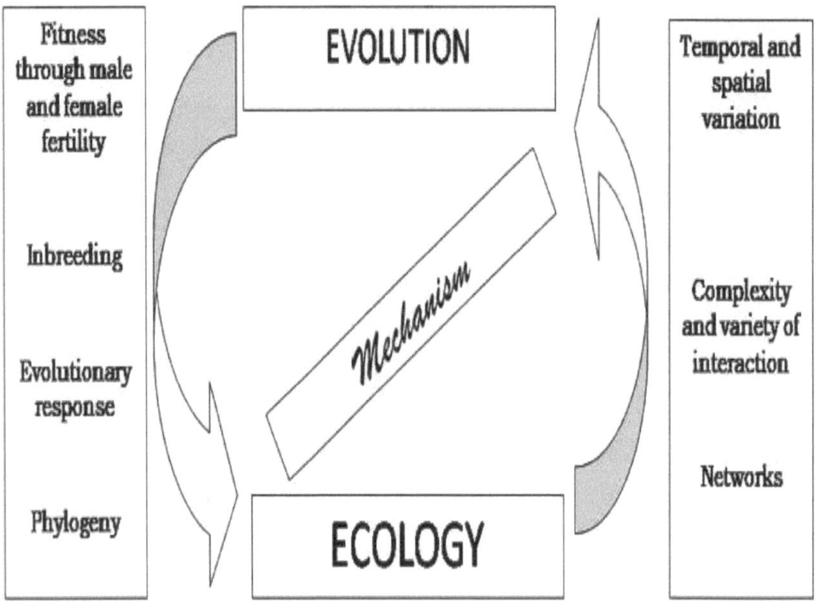

Figure 22.

Conceptual representation of the interplay between ecology and evolution in the study of plant-pollinator interactions.

There are many reasons for which pollinators visit flowers, including feeding, pollen collection, warmth in some cases, oils and resins, as well as for shelter and mating rendezvous sites. These plant and pollinator interactions as mutualisms sustain not only plant diversity, but also the diversity of an estimated 350,000 animal species, including insects, birds, and mammals. Ratto et al. [365] reported an average 63% loss of fruit or seed production when vertebrate pollinators are excluded from the flowering plants' ecology they visit. These results often reported experimentally that selective exclusion of a single group of an effective pollinators from plant-pollinator interaction can result in the failure of plants to produce fruits or seeds [361- 365].

Diversity of pollinators in habitat can compete for floral resources, the declining pollinators population strengthens existing plant-pollinator interaction or allow new plant-pollinator interaction to form. The diverse pollen feeding behavior by bee species is due to digestibility and nutritional content requirement fulfillment. There are specialized flower plant-pollinator relationships like certain solitary bees species], reduction of these flower plants from habitat often results in the elimination of their specialist plant-pollinator populations. Viana et al. [371] evaluated more than 250 studies that showed the impact of landscape and pollinators interactions.

The forage bees' ability to assess the nutritional value of pollen sources before establishing plant-pollinator interaction is valuable. A recent study on pollination ecology mainly emphasizes three aspects, first ecological (pollination involving one or few kinds of plant and animals), second phenotypic (having specialized flowers or morphologies) and third is evolutionary (showing transitions towards increased specialization) [366- 374].

Declining pollinator, a potential threat

It is difficult to determine as less surveys are organized to record whether pollinator species are declining around the world. If we study the literature many explanations have been invoked to account for declines in pollinator population around the globe. There are a few of these reasons such as exposure to pathogens, parasites, and pesticides; habitat fragmentation and loss; climate change; market forces; intra and inter specific competition with native and invasive species; and genetic alterations. Reduction in pollinator diversity or abundance may influence the amount and source of pollen deposited on the reproductive part of flower or stigma [375- 380].

The western honeybee, *Apis mellifera*, colony losses are elucidated in many literatures with attack of different honey pests i.e. parasites, pathogens [76], mostly predominately identified parasitic mites *Varroa destructor* and *Acarapis woodi*, the pathogen *Paenibacillus larvae* (American foulbrood) and the invasive Africanized honey bee. The application of pesticides, especially insecticides in crops, vegetable, and orchards to control pests, kills or weakens thousands of honey bee colonies and affect their foraging and nesting behaviors that prevent plant pollination. The basic behind pesticides to kill or weaken the colony is the result of accidents, careless application, or failure to adhere to label recommendations and warnings. Some of the advance studies showed that transgenic crops developed to reduce the unintended effects of pesticides have shown that there are direct effects on non-target species, including some pollinators. Transgenic crops that express transgene with varied expression levels have not been yet reported effects on honeybee [381- 395].

There are degradations reported around the globe in habitat i.e. alternations, fragmentation, and loss cause major problems for populations of many organisms, and pollinator populations are also one of them. Insect pollinator‘s i.e. bees and others require nesting sites (suitable soil, dead wood, abandoned mouse nests, and burrows) and floral resources (nectar and pollen) to exist. These habitat resources are at extinction through the disruption caused by row crop agriculture, grazing, and fragmentation of habitat into patches, which are small enough for the survival of diverse communities of pollinators. Some other reported causes of decline in pollinators‘ population are monoculture, the lower density of weed flora, declining pastures, loss of flower-rich grasslands, and overgrazing can disrupt the nesting of bees [396- 401].

Industrial development around the global, regional and local climate changes can alter or disrupt plant-pollinator relationships. Many studies and reports show the climate change forecast is shifted in temperature and precipitation, concentrations of carbon dioxide and ozone, and ultraviolet levels effects pollinators in many ways. There is evidence that the latitudinal and altitudinal ranges of some plants and pollinators have changed in the past 30 years, presumably in response to global warming [402- 408].

Management and restoration of pollinators

Information on the status of most of the pollinators is incomplete around the world, and it is in a natal stage in developing countries. Much can be done to maintain commercial and wild pollinator populations and to prevent future shortages of pollination services. Indigenous communities have an important role in the conservation of habitats through customary laws/rules, these areas are important biodiversity refuges providing valuable ecosystem services including pollination, which improved crop pollination in adjacent farming landscape [409-412].

An agri-environment scheme, on farmlands, has been proactively practiced in European Union countries through incentives to support biodiversity. The US Farm bill (2008) had made specific economic provisions for pollinator conservation when it was further ratified in the 2014 Farm bill. The potential of conserving non-cropped land as a model in agro-ecosystem can be proved vital, through these agri- environment scheme models of conservation pollinators in the agriculture landscape can go a long way to inoculate pollinators naturally. Mostly, pollinators are transported over long distances for the purpose of pollination. They are also transported outside of their natural distribution range (e.g. African honey bees into Brazil, European bumble bees into Australia, Asia, and South America) [413- 416].

Best management practices (BMP), similar to Good Agricultural Practice (GAP) should be promoted by the FAO in apiaries that need to be developed that respect local differences in beekeeping and hive management at the country level. There should be non-compulsive suggestions have been put forward overall lacking international harmonization. In this perspective resistant stock of bees against parasitic mites is to be developed, identify the locally adapted stock of bees, instrumental insemination in bees, selection and managing miticide resistance in bees, etc.[417]

Through collective approaches either for native and introduced bee species, whether solitary or social, requires the correct balance of water, flora hosts that offer sufficient pollen and nectar of the correct types, nest building materials (leaves, resin, sap, gums, floral oils, essential oils, bark, plant trichomes, old mouse nests, snail shells, mud, sand, pebbles), and nesting substrates to survive as adults and rear their larval broods (Table 6) [418- 420].

Table 6.

Pollinators and resources requirements.

S. no.	Pollinator group	Resource function	Resource
1.	Honey bees, bumble bees	Nesting, roosting sites, or substrates	Cavities (underground, hollow trees)
2.	Nonsocial bees, wasps	Nesting sites or substrates	Bare ground, vertical cliffs or ditch bank, adobe walls
3.	Bumble bees	Nesting sites	Rodent, mouse nests
4.	Flies	Adult food	Pollen, nectare
5.	Leafcutter bees, mason bees	Building materials	Leaves cut into pieces or masticated
6.	Orchid bees	Pheromones	Essential oils, such as monoterpenoids collected by males
7.	Ants	Adult, larval food	Nectar, honeydew, insect prey

Limitations that restrict the pollinator-plant interaction

Deforestation and habitat changes have shown an adverse effect on insect pollinators, seed predators, decomposers, and parasitoids, which are highly susceptible to these changes. The success of plant reproduction may be sensitive to the loss of pollinators. Some the evidence suggests that pollinator populations are declining worldwide. These changes result in the destruction of food sources, nesting, oviposition, resting, and mating sites.

The increase in population pressure and urbanization of wild and agricultural lands has disrupted the habitat of wild pollinators viz., moths, butterflies, and bees and managed pollinators experienced sudden colony losses. With the increased demand for food crops and higher productivity by use of either plant production or plant protection chemicals has killed the pollinators directly, and eradicated alternative pollen sources from their natural forage species [421- 429].

The pollination host range of honeybees is wide, but they do not pollinate all types of the crop with equal efficiency, are not active under all climatic conditions. Whereas, some of the bees have the ability to pollinate some crops at a higher level of efficiency, with their lower population densities, and with greater independence of climatic conditions [430, 431].

Database of wild pollinator populations and communities is one of severe lack of long term planning and evaluation of their valuation in much of the world, especially for invertebrate pollinators. Such as European red List of bees, 57% of the European bee species were categorized as ―data deficient"; butterflies and moths from parts of Africa that are described at threatened status also reported in the literature [432-437].

Emerging risks such as diseases, invasive alien species, pathogens, etc., threaten the pollinators, there should be phytosanitary and preventive measures that could be accompanied for the effective response to these emerging risks. Few regions in the world (parts of Australia, Seychelles), that are not affected by the ectoparasitic mite, *Varroa destructor*, the most detrimental honey bee pest. These mites also act as a vector for a number of bee viruses, which might spill over to wild bee species [438- 440].

Conclusion

Our findings demonstrate that a rich variety of flowering plants significantly enhances honey bee foraging activity, which is crucial for sustaining healthy bee populations and improving honey production. Specifically, the presence of both native and non-native flowering species is vital, as they collectively provide a continuous supply of nectar and pollen throughout the year. Additionally, the study shows that honey bees often forage within 0.5 km of their hives, emphasising the importance of having diverse plants and abundant floral resources close to apiaries to maximise the foraging efficiency of the honey bee. Conservation efforts should emphasise protecting wild shrubs and herbaceous species that are important vegetation in semi-arid areas and serve as vital sources of floral resources for pollinators. Thus, afforestation programs in the Dodoma region should focus on increasing plant diversity and conserving wild shrubs and herbs rather than creating monocultures of exotic species.

Beekeepers were highly satisfied and showed neutral attitudes toward extension services. Despite the neutral attitudes toward extension services, a considerable portion indicated a higher adoption of recommended beekeeping practices. Moreover, rural women needed professional guidance and training for beekeeping.

Beekeepers in the young age group seemed less satisfied with the extension services. Highly educated beekeepers were more satisfied with the extension services. The decreased delivery of extension services to beekeepers decreased their attitudes toward extension services.

The findings of the study may have various policy implications. First, policymakers should understand the foremost obstacles to extension services. Initiatives should be put in place to educate the rural women who are involved in beekeeping sectors. Combined efforts should be made by governments, policymakers, and the Ministry of Environment, Water, and Agriculture to initiate training sessions for extension agents to improve the satisfaction and attitudes of beekeepers. Satisfaction and positive attitudes toward extension services could make knowledge transfer easy among beekeepers. Moreover, new research should be conducted to investigate the innovative, practical viability, and potential of sustainable beekeeping practices as an alternative to traditional beekeeping practices. The results of future research should be used for policymaking to stimulate sustainable beekeeping practices at the local level and, in turn, to reach one of the core goals of Vision 2030 for sustainable honey production.

we found that various insects, including different bee species, visited fennel flowers for pollination. Notably, honeybees *Apis mellifera* was the most effective pollinators due to their frequent visits and high pollen loads. The abundant fennel blooms were a great additional food source for honeybees. Hives near fennel fields saw significant improvements in colony health, with more sealed brood, food storage, and overall brood production. Comparing bees' effect on fennel plots, honeybee visits boosted attributes like fruit set, weight, and seed production. These benefits underline the economic value of conserving wild bee populations that pollinate fennel. Our study shows how fennel and honeybees help each other. Fennel provides food for bees, and bee pollination enhances fennel growth. To sustainably produce fennel, we suggest managing both crops and pollinators. Future research can explore landscape management to optimize these mutual bee-crop relationships.

Sustainable agriculture requires reliable pollinators, but a shortage of pollinators could not be strongly evidenced for food crisis or pollination crisis. Long-term data deficient on the pollinators' population should be noted and there is no evidence of their decline over time, neither there is a framed definition to label pollinator crisis universally on that frame. The honeybee is a valuable pollinator to perform an important pollination function in the ecosystem. The decline in a number of managed pollinators in the system is due to some of the reasons such as introduced parasites and pathogens.

There is a need for time to be compatible with and comprehensive management strategy of crop pollination for sustainable agriculture. Pollinators require to be managed through augmentation or conservation as needed to study their biology and ecology. Several studies that show the declining bee population poses a threat to global food security. Nesting habitat must be provided whether as a soil bed of a more or less special nature, or as stumps of trees and logs, or as rodent burrows for bumbles. Conservation of native pollinator habitat can be enhanced by changes in land use management strategies viz., non-cultivated patches of ground, setting up parks or protected areas for wildlife, flora, and fauna both at public and private areas.

There should be a policy for arable, non-arable, and along with the roadside land that could facilitate the planting of wild plant flora which encourages pollinator populations. There should be judicious and timely use of pesticides that should ensure the protection of pollinators. There should be a crop pollination plan for all pollinator-dependent crops that must be included in the national or state crop production strategies. Farmer's awareness camps should be organized in the rural areas about crop pollination and the role of pollinators should be described, so that there may be a change in plant protection chemicals patterns.

References

1. Burkle, L.A.; Marlin, J.C.; Knight, T.M. Plant-pollinator interactions over 120 years: Loss of species, co-occurrence, and function. *Science* **2013**, *339*, 1611–1615.
2. Hallmann, C.A.; Sorg, M.; Jongejans, E.; Siepel, H.; Hofland, N.; Schwan, H.; Stenmans, W.; Müller, A.; Sumser, H.; Hörren, T.; et al. More than 75 percent decline over 27 years in total flying insect biomass in protected areas. *PLoS ONE* **2017**, *12*, e0185809.
3. Hallmann, C.A.; Ssymank, A.; Sorg, M.; de Kroon, H.; Jongejans, E. Insect biomass decline scaled to species diversity: General patterns derived from a hoverfly community. *Proc. Natl. Acad. Sci. USA* **2021**, *118*, e2002554117.
4. Kosior, A.; Celary, W.; Olejniczak, P.; Fijał, J.; Król, W.; Solarz, W.; Płonka, P. The decline of the bumble bees and cuckoo bees (Hymenoptera: Apidae: Bombini) of Western and Central Europe. *Oryx* **2007**, *41*, 79–88.
5. Sánchez-Bayo, F.; Wyckhuys, K.A.G. Worldwide decline of the entomofauna: A review of its drivers. *Biol. Conserv.* **2019**, *232*, 8–27.
6. Carr-Markell, M.K.; Demler, C.M.; Couvillon, M.J.; Schürch, R.; Spivak, M. Do honey bee (*Apis mellifera*) foragers recruit their nestmates to native forbs in reconstructed prairie habitats? *PLoS ONE* **2020**, *15*, e0228169.
7. Albrecht, M.; Kleijn, D.; Williams, N.M.; Tschumi, M.; Blaauw, B.R.; Bommarco, R.; Campbell, A.J.; Dainese, M.; Drummond, F.A.; Entling, M.H. The effectiveness of flower strips and hedgerows on pest control, pollination services and crop yield: A quantitative synthesis. *Ecol. Lett.* **2020**, *23*, 1488–1498.

8. Fitch, G.; Wilson, C.J.; Glaum, P.; Vaidya, C.; Simao, M.-C.; Jamieson, M.A. Does urbanization favour exotic bee species? Implications for the conservation of native bees in cities. *Biol. Lett.* **2019**, *15*, 20190574.
9. Hopwood, J.L. The contribution of roadside grassland restorations to native bee conservation. *Biol. Conserv.* **2008**, *141*, 2632–2640.
10. Meyer, S.; Unternährer, D.; Arlettaz, R.; Humbert, J.-Y.; Menz, M.H. Promoting diverse communities of wild bees and hoverflies requires a landscape approach to managing meadows. *Agric. Ecosyst. Environ.* **2017**, *239*, 376–384.
11. Wilson, C.J.; Jamieson, M.A. The effects of urbanization on bee communities depends on floral resource availability and bee functional traits. *PLoS ONE* **2019**, *14*, e0225852.
12. Winfree, R.; Bartomeus, I.; Cariveau, D.P. Native pollinators in anthropogenic habitats. *Annu. Rev. Ecol. Evol. System.* **2011**, *42*, 1–22.
13. Buri, P.; Arlettaz, R.; Humbert, J.-Y. Delaying mowing and leaving uncut refuges boosts orthopterans in extensively managed meadows: Evidence drawn from field-scale experimentation. *Agric. Ecosyst. Environ.* **2013**, *181*, 22–30.
14. Buri, P.; Humbert, J.-Y.; Arlettaz, R. Promoting pollinating insects in intensive agricultural matrices: Field-scale experimental manipulation of hay-meadow mowing regimes and its effects on bees. *PLoS ONE* **2014**, *9*, e85635.
15. Larson, J.L.; Kesheimer, A.J.; Potter, D.A. Pollinator assemblages on dandelions and white clover in urban and suburban lawns. *J. Insect Conserv.* **2014**, *18*, 863–873.

16. Garbuzov, M.; Fensome, K.A.; Ratnieks, F.L.W. Public approval plus more wildlife: Twin benefits of reduced mowing of amenity grass in a suburban public park in Saltdean, UK. *Insect Conserv. Divers.* **2015**, *8*, 107–119.
17. Campbell, B.; Khachatryan, H.; Rihn, A. Pollinator-friendly plants: Reasons for and barriers to purchase. *HortTechnology* **2017**, *27*, 831–839.
18. Kalaman, H.; Knox, G.W.; Wilson, S.B.; Wilber, W. A master gardener survey: Promoting pollinator-friendly plants through education and outreach. *HortTechnology* **2020**, *30*, 163–167.
19. Goddard, M.A.; Dougill, A.J.; Benton, T.G. Scaling up from gardens: Biodiversity conservation in urban environments. *Trends Ecol. Evol.* **2010**, *25*, 90–98.
20. Davies, Z.G.; Fuller, R.A.; Loram, A.; Irvine, K.N.; Sims, V.; Gaston, K.J. A national scale inventory of resource provision for biodiversity within domestic gardens. *Biol. Conserv.* **2009**, *142*, 761–771.
21. Taylor, L. From ways of life to lifestyle: Theordinari-ization'of British gardening lifestyle television. *Eur. J. Commun.* **2002**, *17*, 479–493.
22. Kaluza, B.F.; Wallace, H.; Heard, T.A.; Klein, A.M.; Leonhardt, S.D. Urban gardens promote bee foraging over natural habitats and plantations. *Ecol. Evol.* **2016**, *6*, 1304–1316.]
23. Baldock, K.C.; Goddard, M.A.; Hicks, D.M.; Kunin, W.E.; Mitschunas, N.; Morse, H.; Osgathorpe, L.M.; Potts, S.G.; Robertson, K.M.; Scott, A.V. A systems approach reveals urban pollinator hotspots and conservation opportunities. *Nat. Ecol. Evol.* **2019**, *3*, 363–373.
24. Lanner, J.; Kratschmer, S.; Petrović, B.; Gaulhofer, F.; Meimberg, H.; Pachinger, B. City dwelling wild bees: How communal gardens promote species richness. *Urban Ecosyst.* **2020**, *23*, 271–288. Erickson, E.; Adam, S.;

Russo, L.; Wojcik, V.; Patch, H.; Grozinger, C. More than meets the eye? The role of annual ornamental flowers in supporting pollinators. *Environ. Entomol.* **2020**, *49*, 178–188.

25. Garbuzov, M.; Samuelson, E.E.W.; Ratnieks, F.L.W. Survey of insect visitation of ornamental flowers in Southover Grange garden, Lewes, UK. *Insect Sci.* **2015**, *22*, 700–705.
26. Garbuzov, M.; Ratnieks, F.L. Using the British national collection of asters to compare the attractiveness of 228 varieties to flower-visiting insects. *Environ. Entomol.* **2015**, *44*, 638–646.
27. Harris, B.A.; Braman, S.K.; Pennisi, S.V. Influence of plant taxa on pollinator, butterfly, and beneficial insect visitation. *HortScience* **2016**, *51*, 1016–1019.
28. Garbuzov, M.; Ratnieks, F.L. Listmania: The strengths and weaknesses of lists of garden plants to help pollinators. *BioScience* **2014**, *64*, 1019–1026.
29. Garbuzov, M.; Ratnieks, F.L. Quantifying variation among garden plants in attractiveness to bees and other flower-visiting insects. *Funct. Ecol.* **2014**, *28*, 364–374.
30. Rollings, R.; Goulson, D. Quantifying the attractiveness of garden flowers for pollinators. *J. Insect Conserv.* **2019**, *23*, 803–817.
31. Erickson, E.; Patch, H.; Grozinger, C. Herbaceous perennial ornamental plants can support complex pollinator communities. *Sci. Rep.* **2021**, *11*, 17352.
32. Ratnieks, F.L.W.; Shackleton, K. Does the Waggle Dance Help Honey Bees to Forage at Greater Distances than Expected for their Body Size? *Front. Ecol. Evol.* **2015**, *3*, 31.

33. DiSabato-Aust, T. *The Well-Tended Perennial Garden: The Essential Guide to Planting + Pruning Techniques*, 3rd ed.; Timber Press, Inc.: Portland, OR, USA, 2017.
34. Adamson, N.L.; Borders, B.; Cruz, J.K.; Jordan, S.F.; Gill, K.; Hopwood, J.; Lee-Mader, E.; Minnerath, A.; Vaughan, M. *Pollinator Plants: Mid-Atlantic Region*; Xerces Society: Portland, OR, USA, 2017.
35. Daniel, J.B.; Glennon, B.; Gordon, S.; Wentz, C. *Virginia Plant Establishment Guide*; USDA-NRCS: Washington, DC, USA, 2011.
36. Garbuzov, M.; Madsen, A.; Ratnieks, F.L. Patch size has no effect on insect visitation rate per unit area in garden-scale flower patches. *Acta Oecologica* **2015**, *62*, 53–57.
37. R Core Team. *R: A Language and Environment for Statistical Computing*; R Foundation for Statistical Computing: Vienna, Austria, 2020.
38. Bates, D.; Mächler, M.; Bolker, B.; Walker, S. Fitting Linear Mixed-Effects Models Using lme4. *J. Stat. Softw.* **2015**, *67*, 48.
39. Morris, E.K.; Caruso, T.; Buscot, F.; Fischer, M.; Hancock, C.; Maier, T.S.; Meiners, T.; Müller, C.; Obermaier, E.; Prati, D.; et al. Choosing and using diversity indices: Insights for ecological applications from the German Biodiversity Exploratories. *Ecol. Evol.* **2014**, *4*, 3514–3524.
40. Hutcheson, K. A test for comparing diversities based on the Shannon formula. *J. Theor. Biol.* **1970**, *29*, 151–154.
41. Barr, D.J.; Levy, R.; Scheepers, C.; Tily, H.J. Random effects structure for confirmatory hypothesis testing: Keep it maximal. *J. Mem. Lang.* **2013**, *68*, 255–278.
42. Lenth, R.V. Emmeans: Estimated Marginal Means, Aka Least Squares Means. Available online: **https://cran.r-**

project.org/web/packages/emmeans/index.html (accessed on 25 September 2022).
43. Wardhaugh, C.W. How many species of arthropods visit flowers? *Arthropod-Plant Interact.* **2015**, *9*, 547–565.
44. Ollerton, J. *Pollinators and Pollination: Nature and Society*; Pelagic Publishing Ltd.: Exeter, UK, 2021.
45. Klein, A.-M.; Vaissière, B.E.; Cane, J.H.; Steffan-Dewenter, I.; Cunningham, S.A.; Kremen, C.; Tscharntke, T. Importance of pollinators in changing landscapes for world crops. *Proc. Biol. Sci.* **2007**, *274*, 303–313.
46. Colla, S.R.; Packer, L. Evidence for decline in eastern North American bumblebees (Hymenoptera: Apidae), with special focus on *Bombus affinis* Cresson. *Biodivers. Conserv.* **2008**, *17*, 1379–1391.
47. Theodorou, P.; Albig, K.; Radzevičiūtė, R.; Settele, J.; Schweiger, O.; Murray, T.E.; Paxton, R.J. The structure of flower visitor networks in relation to pollination across an agricultural to urban gradient. *Funct. Ecol.* **2017**, *31*, 838–847.
48. Goulson, D.; Hughes, W.; Derwent, L.; Stout, J. Colony growth of the bumblebee, Bombus terrestris, in improved and conventional agricultural and suburban habitats. *Oecologia* **2002**, *130*, 267–273.
49. Pardee, G.L.; Philpott, S.M. Native plants are the bee's knees: Local and landscape predictors of bee richness and abundance in backyard gardens. *Urban Ecosyst.* **2014**, *17*, 641–659.
50. Fukase, J.; Simons, A. Increased pollinator activity in urban gardens with more native flora. *Appl. Ecol. Environ. Res.* **2016**, *14*, 297–310.

51. Lowenstein, D.M.; Matteson, K.C.; Minor, E.S. Evaluating the dependence of urban pollinators on ornamental, non-native, and ‗weedy' floral resources. *Urban Ecosyst.* **2019**, *22*, 293–302.
52. Kalaman, H.; Wilson, S.B.; Mallinger, R.E.; Knox, G.W.; Kim, T.; Begcy, K.; van Santen, E. Evaluation of Native and Nonnative Ornamentals as Pollinator Plants in Florida: II. Floral Resource Value. *HortScience* **2022**, *57*, 137–143.
53. Matteson, K.C.; Langellotto, G.A. Small scale additions of native plants fail to increase beneficial insect richness in urban gardens. *Insect Conserv. Divers.* **2011**, *4*, 89–98.
54. Garbuzov, M.; Alton, K.; Ratnieks, F.L. Most ornamental plants on sale in garden centres are unattractive to flower-visiting insects. *PeerJ* **2017**, *5*, e3066.
55. Steffan-Dewenter, I.; Tscharntke, T. Succession of bee communities on fallows. *Ecography* **2001**, *24*, 83–93.
56. Fussell, M.; Corbet, S.A. Flower usage by bumble-bees: A basis for forage plant management. *J. Appl. Ecol.* **1992**, *29*, 451–465.
57. Parrish, J.; Bazzaz, F. Difference in pollination niche relationships in early and late successional plant communities. *Ecology* **1979**, *60*, 597–610.
58. Corbet, S.A. Insects, plants and succession: Advantages of long-term set-aside. *Agric. Ecosyst. Environ.* **1995**, *53*, 201–217.
59. Willmer, P. *Pollination and Floral Ecology*; Princeton University Press: Princeton, NJ, USA, 2011.
60. Tuell, J.K.; Fiedler, A.K.; Landis, D.; Isaacs, R. Visitation by wild and managed bees (Hymenoptera: Apoidea) to eastern US native plants for use in conservation programs. *Environ. Entomol.* **2014**, *37*, 707–718.

61. Weisenhorn, J. *Annual Flowers That Attract Pollinators*; University of Minnesota Extension: Minneapolis, MN, USA, 2021.
62. Božanić Tanjga, B.; Ljubojević, M.; Đukić, A.; Vukosavljev, M.; Ilić, O.; Narandžić, T. Selection of Garden Roses to Improve the Ecosystem Services They Provide. *Horticulturae* **2022**, *8*, 883.
63. Strzałkowska-Abramek, M. Nectar and pollen production in ornamental cultivars of *Prunus serrulata* (Rosaceae). *Folia Hortic.* **2019**, *31*, 205–212.
64. Hicks, D.M.; Ouvrard, P.; Baldock, K.C.; Baude, M.; Goddard, M.A.; Kunin, W.E.; Mitschunas, N.; Memmott, J.; Morse, H.; Nikolitsi, M. Food for pollinators: Quantifying the nectar and pollen resources of urban flower meadows. *PLoS ONE* **2016**, *11*, e0158117.
65. Staab, M.; Pereira-Peixoto, M.H.; Klein, A.-M. Exotic garden plants partly substitute for native plants as resources for pollinators when native plants become seasonally scarce. *Oecologia* **2020**, *194*, 465–480.
66. Cubey, J.; Merrick, J. *RHS Plant Finder 2022*; Royal Horticultural Society: London, UK, 2022.
67. NRCS USDA. *The PLANTS Database*; United States Department of Agriculture: Washington, DC, USA, 2022.
68. Biesmeijer, J.C.; Slaa, J.E. The structure of eusocial bee assemblages in Brazil. *Apidologie* **2006**, *37*, 240–258.
69. von Frisch, K. *The Dance Language and Orientation of Bees*; Harvard University Press: Cambridge, MA, USA, 1967.
70. Couvillon, M.J. The dance legacy of Karl von Frisch. *Insect. Soc.* **2012**, *59*, 297–306
71. Garbuzov, M.; Couvillon, M.J.; Schürch, R.; Ratnieks, F.L.W. Honey bee dance decoding and pollen-load analysis shows limited foraging on oilseed

rape—A potential source of neonicotinoid contamination. *Agric. Ecosyst. Environ.* **2015**, *203*, 62–68.

72. Balfour, N.J.; Ratnieks, F.L. Using the waggle dance to determine the spatial ecology of honey bees during commercial crop pollination. *Agric. For. Entomol.* **2017**, *19*, 210–216.

73. Couvillon, M.J.; Ratnieks, F.L.W. Environmental consultancy: Dancing bee bioindicators to evaluate landscape "health". *Front. Ecol. Evol.* **2015**, *3*.

74. Couvillon, M.J.; Schürch, R.; Ratnieks, F.L.W. Dancing bees communicate a foraging preference for rural lands under High Level Agri-Environment Schemes. *Curr. Biol.* **2014**, *24*, 1212–1215.

75. Mallinger, R.E.; Gaines-Day, H.R.; Gratton, C. Do managed bees have negative effects on wild bees?: A systematic review of the literature. *PLoS ONE* **2017**, *12*, e0189268.

76. Paini, D. Impact of the introduced honey bee (*Apis mellifera*)(Hymenoptera: Apidae) on native bees: A review. *Austral Ecol.* **2004**, *29*, 399–407.

77. Goulson, D. Effects of introduced bees on native ecosystems. *Annu. Rev. Ecol. Evol. Syst.* **2003**, *34*, 1–26.

78. Angelella, G.M.; McCullough, C.T.; O'Rourke, M.E. Honey bee hives decrease wild bee abundance, species richness, and fruit count on farms regardless of wildflower strips. *Sci. Rep.* **2021**, *11*, 3202.

79. Balfour, N.J.; Garbuzov, M.; Ratnieks, F.L.W. Longer tongues and swifter handling: Why do more bumble bees (*Bombus* spp.) than honey bees (*Apis mellifera*) forage on lavender (*Lavandula* spp.). *Ecol. Entomol.* **2013**, *38*, 323–329.

80. Couvillon, M.J.; Schürch, R.; Ratnieks, F.L.W. Waggle Dance Distances as Integrative Indicators of Seasonal Foraging Challenges. *PLoS ONE* **2014**, *9*, e93495.
81. Garbuzov, M.; Schürch, R.; Ratnieks, F.W. Eating locally: Dance decoding demonstrates that urban honey bees in Brighton, UK, forage mainly in the surrounding urban area. *Urban Ecosyst.* **2014**, *18*, 411–418.
82. Sponsler, D.B.; Matcham, E.G.; Lin, C.-H.; Lanterman, J.L.; Johnson, R.M. Spatial and taxonomic patterns of honey bee foraging: A choice test between urban and agricultural landscapes. *J. Urban Ecol.* **2017**, *3*, juw008.
83. Couvillon, M.J.; Fensome, K.A.; Quah, S.K.L.; Schürch, R. Summertime blues: August foraging leaves honey bees empty-handed. *Commun. Integr. Biol.* **2014**, *7*, e28821.
84. Steele, T.N.; Schürch, R.; Ohlinger, B.D.; Couvillon, M.J. Apple orchards feed honey bees during, but even more so after, bloom. *Ecosphere* **2022**, *13*, e4228.
85. Erickson, E.; Junker, R.; Ali, J.; McCartney, N.; Patch, H.; Grozinger, C. Complex floral traits shape pollinator attraction to ornamental plants. *Ann. Bot.* **2022**, *130*, 561–577.
86. Marquardt, M.; Kienbaum, L.; Kretschmer, L.A.; Penell, A.; Schweikert, K.; Ruttensperger, U.; Rosenkranz, P. Evaluation of the importance of ornamental plants for pollinators in urban and suburban areas in Stuttgart, Germany. *Urban Ecosyst.* **2021**, *24*, 811–825.
87. WorldAtlas Which Crops and Plants Are Pollinated by Honeybees? [(accessed on 1 June 2023)]. Available online: https://www.worldatlas.com/articles/which-crops-plants-are-pollinated-by-honey-bees.html.

88. Stanley D.A., Msweli S., Johnson S.D. Native honeybees as flower visitors and pollinators in wild plant communities in a biodiversity hotspot. Ecosphere. 2020;11:e02957. doi: 10.1002/ecs2.2957.

89. Proctor M., Yeo P., Lack A. The Natural History of Pollination. Collins; UK, London: 2009.

90. Sundar P.S., Chowdhury C., Kamarthi S. Analysis of Design of Flowers and Honeybees in the Pollination Process; Proceedings of the 4th Asia Pacific Conference on Industrial Engineering and Operations Management; Ho Chi Minh City, Vietnam. 12–14 September 2023.

91. Suh N.P. Axiomatic Design: Advances and Applications. Oxford University Press; New York, NY, USA: 2001. pp. 11–20. Chapters 1.7.1–1.7.4.

92. Suh N.P. Axiomatic Design: Advances and Applications. Oxford University Press; New York, NY, USA: 2001. pp. 39–50. Chapters 1.7.9–1.7.11.

93. Quirós G. Honey Bees Make Honey … and Bread? 2019. [(accessed on 30 September 2023)]. Available online: https://www.kqed.org/science/1940898/honey-bees-make-honey-and-bread.

94. van der Kooi C.J., Vallejo-Marín M., Leonhardt S.D. Mutualisms and (A)symmetry in Plant–Pollinator Interactions. Curr. Biol. 2021;31:R91–R99. doi: 10.1016/j.cub.2020.11.020.

95. Hung K.-L.J., Kingston J.M., Albrecht M., Holway D.A., Kohn J.R. The worldwide importance of honeybees as pollinators in natural habitats. Proc. R. Soc. B. 2018;285:20172140. doi: 10.1098/rspb.2017.2140.

96. Chambó E., Garcia R., Oliveira N., Duarte-Júnior J. Honeybee Visitation to Sunflower: Effects on Pollination and Plant Genotype. Sci. Agric. 2021;68:647–651. doi: 10.1590/S0103-90162011000600007.

97. 11.Willmer P. Pollination and Floral Ecology. Princeton University Press; Princeton, NJ, USA: 2011. pp. 378–417.

98. The Basitarsal Brushes of Bees Are Used to Clean the Pollen off Their Faces While They Forage. [(accessed on 30 September 2023)]. Available online: https://asknature.org/strategy/brushes-clean-off-pollen/

99. Horridge A. The Discovery of a Visual System: The Honeybee. CABI; Oxfordshire, UK: 2019. pp. 197–221.

100. Dyer F.C., Gould J.L. Honey Bee Navigation: The honey bee‘s ability to find its way depends on a hierarchy of sophisticated orientation mechanisms. Am. Sci. 1983;71:587–597.

101. Ribi W., Warrant E., Zeil J. The organization of honeybee ocelli: Regional specializations and rhabdom arrangements. Arthropod Struct. Dev. 2011;40:509–520. doi: 10.1016/j.asd.2011.06.004.

102. Collett T.S. Insect Navigation: Visual Panoramas and the Sky Compass. Curr. Biol. 2008;18:R1058–R1061. doi: 10.1016/j.cub.2008.10.006.

103. Dacke M., Srinivasan M. Two odometers in honeybees? J. Exp. Biol. 2008;211:3281–3286. doi: 10.1242/jeb.021022.

104. Najera D.A., McCullough E.L., Jander R. Honeybees Use Celestial and/or Terrestrial Compass Cues for Inter-Patch Navigation. Ethology. 2014;121:94–102. doi: 10.1111/eth.12319.

105. Dacke M., Srinivasan M.V. Honeybee navigation: Distance estimation in the third dimension. J. Exp. Biol. 2007;210:845–853. doi: 10.1242/jeb.002089.

106. Eberhard W.G., Wcislo W.T. Plenty of Room at the Bottom? Tiny animals solve problems of housing and maintaining oversized brains, shedding new light on nervous-system evolution. Am. Sci. 2012;100:226–233. doi: 10.1511/2012.96.226.

107. Chittka L. The Mind of a Bee. Princeton University Press; Princeton, NJ, USA: 2022. pp. 143–164.
108. Rueppell O., Bachelier C., Fondrk M.K., Page R.E., Jr. Regulation of life history determines lifespan of worker honeybees (Apis mellifera L.) Exp. Gerontol. 2007;42:1020–1032. doi: 10.1016/j.exger.2007.06.002.
109. Kirchner W.H., Towne W.F. The Sensory Basis of the Honeybee‘s Dance Language. Sci. Am. 1994;270:74–80. doi: 10.1038/scientificamerican0694-74.
110. Nobel Foundation Archive. [(accessed on 1 June 2023)]. Available online: https://www.nobelprize.org/prizes/medicine/1973/frisch/facts/
111. von Frisch K. The Dance Language and Orientation of Bees. Harvard University Press; Boston, MA, USA: 2014.
112. von Frisch K. The Dancing Bees—An Account of the Life and Senses of the Honeybee. Springer; Berlin, Germany: 1953.
113. Wario F., Wild B., Couvillon M.J., Rojas R., Landgraf T. Automatic methods for long-term tracking and the detection and decoding of communication dances in honeybees. Front. Ecol. Evol. 2015;3:103. doi: 10.3389/fevo.2015.00103.
114. Bortolotti L., Costa C. In: Chemical Communication in the Honey Bee Societyin Neurobiology of Chemical Communication. Mucignat-Caretta C., editor. CRC Press; Boca Raton, FL, USA: 2014. pp. 147–210.
115. Seely T. The Lives of Bees—The Untold Story of the Honeybee in the Wild. Princeton University Press; Princeton, NJ, USA: 2019. pp. 187–214.
116. Wright G.A., Lutmerding A., Dudareva N., Smith B.H. Intensity and the ratios of compounds in the scent of snapdragon flowers affect scent

discrimination by honeybees (Apis mellifera) J. Comp. Physiol. A. 2005;191:105–114. doi: 10.1007/s00359-004-0576-6.

117. Dinkel T., Lunau K. How drone flies (Eristalis tenax L., Syrphidae, Diptera) use floral guides to locate food sources. J. Insect Physiol. 2001;47:1111–1118. doi: 10.1016/S0022-1910(01)00080-4.

118. Clarke D., Morley E., Robert D. The bee, the flower, and the electric field: Electric ecology and aerial electroreception. J. Comp. Physiol. A. 2017;203:737–748. doi: 10.1007/s00359-017-1176-6.

119. Vaknin Y., Gan-Mor S., Bechar A., Ronen B., Eisikowitch D. The role of electrostatic forces in pollination. In: Dafni A., Hesse M., Pacini E., editors. Pollen and Pollination. Springer; Vienna, Austria: 2000. pp. 133–142.

120. Futuyma D.J., Kirkpatrick M. Evolution. Sinauer Associates; Sunderland, MA, USA: 2017. p. 322.

121. Friis E.M., Crane P.R., Pedersen K.R. Early Flowers and Angiosperm Evolution. Cambridge University Press; Cambridge, UK: 2011.

122. O'Toole C., Raw A. Bees of the World—Honeybees. Facts on File Inc.; New York, NY, USA: 2004. pp. 128–142.

123. Suh N.P. Axiomatic Design: Advances and Applications. Oxford University Press; New York, NY, USA: 2001. pp. 21–32. Chapters 1.7.4–1.7.7.

124. Hoffman G., Lande C., Rao S. A Novel Pollen Transfer Mechanism by Honey Bee Foragers on Highbush Blueberry (Ericales: Ericaceae) Environ. Entomol. 2018;47:1465–1470. doi: 10.1093/ee/nvy162.

125. Way R.D. Pollination and Fruit Set of Fruit Crops. Cornell Cooperative Extension Information Bulletin 237. Cornell University. 1995. [(accessed on 1 June 2023)]. Available

online: www.uvm.edu/~orchard/fruit/treefruit/tf_horticulture/AppleHortBasics/Readings/Pollination+and+Fruit+Set+of+Fruit+Crops.pdf.

126. Saccardi L., Brümmer F., Schiebl J., Schwarz O., Kovalev A., Gorb S. Interaction between honeybee mandibles and propolis. Beilstein J. Nanotechnol. 2022;13:958–974. doi: 10.3762/bjnano.13.84.

127. Kernbach S., Thenius R., Kernbach O., Schmickl T. Re-Embodiment of Honeybee Aggregation Behavior in Artificial Micro-Robotic System. Adapt. Behav. 2009;17:237–259. doi: 10.1177/1059712309104966.

128. Braungart M., McDonough W. Cradle to Cradle: Remaking the Way We Make Things. 1st ed. North Point Press; Berkeley, CA, USA: 2002. pp. 92–118.

129. Braungart M., McDonough W. The Upcycle: Beyond Sustainability—Designing for Abundance. 1st ed. North Point Press; Berkeley, CA, USA: 2013. pp. 23–49.

130. Adisorn T., Tholen L., Götz T. Towards a Digital Product Passport Fit for Contributing to a Circular Economy. Energies. 2021;14:2289. doi: 10.3390/en14082289.

131. Ahmed A., Arya S., Gupta V., Furukawa H., Khosla A. 4D printing: Fundamentals, materials, applications and challenges. Polymer. 2021;228:123926. doi: 10.1016/j.polymer.2021.123926.

132. Aldawood F.K. A Comprehensive Review of 4D Printing: State of the Arts, Opportunities, and Challenges. Actuators. 2023;12:101. doi: 10.3390/act12030101.

133. Stürzl W., Boeddeker N., Dittmar L., Egelhaaf M. Mimicking honeybee eyes with a 280° field of view catadioptric imaging system. Bioinspiration Biomim. 2010;5:036002. doi: 10.1088/1748-3182/5/3/036002.

134. Shi G., Wu J., Yan S. Drag Reduction in a Natural High-Frequency Swinging Micro-Articulation: Mouthparts of the Honeybee. J. Insect Sci. 2017;17:30. doi: 10.1093/jisesa/iew122.

135. Hung KL, Kingston JM, Albrecht M, Holway DA, Kohn JR. The worldwide importance of honey bees as pollinators in natural habitats. Proc R Soc B Biol Sci. 1870;2018(285):20172140.

136. Potts SG, Biesmeijer JC, Kremen C, Neumann P, Schweiger O, Kunin WE. Global pollinator declines: trends, impacts and drivers. Trends Ecol Evol. 2010;25(6):345–53.

137. Pires CSS, Maués MM. Insect pollinators, major threats and mitigation measures. Neotrop Entomol. 2020;49:469–71.

138. Nicolson SW. Bee food: the chemistry and nutritional value of nectar, pollen and mixtures of the two. Afr Zool. 2011;46(2):197–204.

139. Ghosh S, Jeon H, Jung C. Foraging behaviour and preference of pollen sources by honey bee (*Apis mellifera*) relative to protein contents. J Ecol Environ. 2020;44:1–7.

140. Hendriksma HP, Oxman KL, Shafir S. Amino acid and carbohydrate tradeoffs by honey bee nectar foragers and their implications for plant–pollinator interactions. J Insect Physiol. 2014;69:56–64.

141. Hoover SE, Ovinge LP. Pollen collection, honey production, and pollination services: managing honey bees in an agricultural setting. J Econ Entomol. 2018;111(4):1509–16.

142. Venjakob C, Ruedenauer FA, Klein AM, Leonhardt SD. Variation in nectar quality across 34 grassland plant species. Plant Biol. 2022;24(1):134–44.

143. da Silva Alencar CL, Nogueira A, Vicente RE, Cotta Coutinho ÍA. Plant species with larger extrafloral nectaries produce better quality nectar when

needed and interact with the best ant partners. J Exp Bot. 2023;74(15):4613–27.

144. Somme L, Vanderplanck M, Michez D, Lombaerde I, Moerman R, Wathelet B, et al. Pollen and nectar quality drive the major and minor floral choices of bumble bees. Apidologie. 2015;46:92–106.

145. Kavitha S, Reddy PR. Influence of floral nectar quantity and quality on honey bee foraging activity in onion. J Entomol Zool Stud. 2019;7(3):92–6.

146. Nicolson SW. Sweet solutions: nectar chemistry and quality. Philos Trans R Soc B. 1853;2022(377):20210163.

147. Nicolson SW, Human H, Pirk CW. Honey bees save energy in honey processing by dehydrating nectar before returning to the nest. Sci Rep. 2022;12(1):16224.

148. Kammerer MA, Biddinger DJ, Rajotte EG, Mortensen DA. Local plant diversity across multiple habitats supports a diverse wild bee community in Pennsylvania apple orchards. Environ Entomol. 2016;45(1):32–8.

149. Decourtye A, Mader E, Desneux N. Landscape enhancement of floral resources for honey bees in agro-ecosystems. Apidologie. 2010;41(3):264–77.

150. Reisman-Berman O, Keasar T, Tel-Zur N. Native and non-native species for dryland afforestation: bridging ecosystem integrity and livelihood support. Ann For Sci. 2019;76(4):1–13.

151. Seitz N, vanEngelsdorp D, Leonhardt SD. Are native and non-native pollinator friendly plants equally valuable for native wild bee communities? Ecol Evol. 2020;10(23):12838–50.

152. Fukase J, Simons AM. Increased pollinator activity in urban gardens with more native flora. Appl Ecol Environ Res. 2016;14(1):297–310.

153. Blaauw BR, Isaacs R. Larger patches of diverse floral resources increase insect pollinator density, diversity, and their pollination of native wildflowers. Basic Appl Ecol. 2014;15(8):701–11.

154. Williams NM, Ward KL, Pope N, Isaacs R, Wilson J, May EA, et al. Native wildflower plantings support wild bee abundance and diversity in agricultural landscapes across the United States. Ecol Appl. 2015;25(8):2119–31.

155. Zhang Y, Tariq A, Hughes AC, Hong D, Wei F, Sun H, et al. Challenges and solutions to biodiversity conservation in arid lands. Sci Total Environ. 2023;857: 159695.

156. Gayo L. Influence of afforestation on coleopterans abundance and diversity at the University of Dodoma. Tanzania Environ Sustain Indic. 2022;16: 100208.

157. Ngongolo K, Kilonzo M. Communities' awareness of afforestation and its contribution to the conservation of lizards in Dodoma, Tanzania. Sci Rep. 2022;12(1):22592.

158. Bremer LL, Farley KA. Does plantation forestry restore biodiversity or create green deserts? A synthesis of the effects of land-use transitions on plant species richness. Biodivers Conserv. 2010;19:3893–915.

159. Athumani PC, Munishi LK, Ngondya IB. Reconstructing Historical Distribution of Large Mammals and their Habitat to Inform Rewilding and Restoration in Central Tanzania. Trop Conserv Sci. 2023;16:1–15.

160. Kayombo CJ, Rubanza C, Giliba RA, Kashindye A. The woody plant species diversity, composition and dominance of Mahungu Green Belt Forest reserve (MGBFR) in Dodoma City, Central Tanzania. East African J Environ Nat Resour. 2020;2(1):1–13.

161. Luswaga H. Carbon sink and shade potential of Trichilia emetica in semi-arid region of Tanzania. Global Ecol Conserv. 2023;45: e02515.

162. Moore E, van Dijk T, Asenga A, Bongers F, Sambalino F, Veenendaal E, et al. Species selection and management under farmer managed natural regeneration in Dodoma, Tanzania. Front For Global Change. 2020;3: 563364.

163. Myeya HE. Recent temperature and rainfall characteristics in Dodoma Region, central Tanzania (1961–2013). Ghana J Geogr. 2021;13(1):63–80.

164. Lasway JV, Peters MK, Njovu HK, Eardley C, Pauly A, Steffan-Dewenter I. Agricultural intensification with seasonal fallow land promotes high bee diversity in Afrotropical drylands. J Appl Ecol. 2022;59(12):3014–26.

165. Dharani N. Field guide to common trees & shrubs of East Africa. South Africa: Penguin Random House; 2011.

166. Van Wyk B. Field guide to trees of southern Africa. South Africa: Penguin Random House; 2013.

167. Fourqurean J, Johnson B, Kauffman JB, Kennedy H, Lovelock C, Saintilan N, et al., Field sampling of vegetative carbon pools in coastal ecosystems. Conservation International, Intergovernmental Oceanographic Commission of UNESCO, International Union for Conservation of Nature: Arlington, VA, USA 2014:67–108.

168. Mcullagh P, Nelder JA, editors. Binary data, generalized linear models. Springer; 1989. p. 98–148.

169. Zuur AF, Ieno EN, Walker NJ, Saveliev AA, Smith GM. Mixed effects models and extensions in ecology with R. New York: Springer; 2009.

170. Borges LA, Sobrinho MS, Lopes AV. Phenology, pollination, and breeding system of the threatened tree *Caesalpinia echinata* Lam.(Fabaceae), and a

review of studies on the reproductive biology in the genus. Flora-Morphol Distrib Funct Ecol Plants. 2009;204(2):111–30.

171. Osman S, Khalek S, Koheil M, El-Haddad A, Wink M. A new steroidal compound (β-sitosterol-3-O-butyl) isolated from *Caesalpinia gilliesii* flowers. Int J Appl Res Nat Prod. 2015;8(2):14–9.

172. Fohouo FNT, Djongwangwe D, Mbianda AP, Messi J, Brückner D. Exploitation of *Dichrostachys cinerea*, *Vitellaria paradoxa, Persea americana* and *Securidaca longepedunculata* flowers by *Apis mellifera adansonii* Latreille (Hymenoptera: Apidae) at Dang (Ngaoundéré, Cameroon). Int J Trop Insect Sci. 2008;28(4):225–33.

173. Krog M, Theilade I. HANSEN H, Ruffo C, Estimating use-values and relative importance of trees to the Kaguru people in semi-arid Tanzania. For Trees Livelihoods. 2005;15(1):25–40.

174. Nicholls CI, Altieri MA. Plant biodiversity enhances bees and other insect pollinators in agroecosystems. A review. Agron Sustain Dev. 2013;33:257–74.

175. Isbell F, Adler PR, Eisenhauer N, Fornara D, Kimmel K, Kremen C, et al. Benefits of increasing plant diversity in sustainable agroecosystems. J Ecol. 2017;105(4):871–9.

176. Abbate A, Campbell JW, Kimmel CB, Kern WH Jr. Urban development decreases bee abundance and diversity within coastal dune systems. Global Ecol Conserv. 2019;20: e00711.

177. Liu CLC, Kuchma O, Krutovsky KV. Mixed-species versus monocultures in plantation forestry: development, benefits, ecosystem services and perspectives for the future. Global Ecol Conserv. 2018;15: e00419.

178. St. Clair AL, Zhang G, Dolezal AG, O'Neal ME, Toth AL. Diversified farming in a monoculture landscape: effects on honey bee health and wild bee communities. Environ Entomol. 2020;49(3):753–64.
179. Heller S, Joshi NK, Leslie T, Rajotte EG, Biddinger DJ. Diversified floral resource plantings support bee communities after apple bloom in commercial orchards. Sci Rep. 2019;9(1):17232.
180. Ebeling A, Klein AM, Schumacher J, Weisser WW, Tscharntke T. How does plant richness affect pollinator richness and temporal stability of flower visits? Oikos. 2008;117(12):1808–15.
181. Morandin LA, Kremen C. Bee preference for native versus exotic plants in restored agricultural hedgerows. Restor Ecol. 2013;21(1):26–32.
182. Rahimi E, Barghjelveh S, Dong P. A review of diversity of bees, the attractiveness of host plants and the effects of landscape variables on bees in urban gardens. Agric Food Secur. 2022;11(1):6.
183. Lu Y, Ranjitkar S, Harrison RD, Xu J, Ou X, Ma X, et al. Selection of native tree species for subtropical forest restoration in Southwest China. PLoS ONE. 2017;12(1): e0170418.
184. Yang Y, Ignatieva M, Gaynor A, Chen C. Urban biodiversity in design: insights into the debate on native versus non-native plants and bees in Western Australia. Urban For Urban Green. 2024;98: 128391.
185. Koyama A, Egawa C, Taki H, Yasuda M, Kanzaki N, Ide T, et al. Non-native plants are a seasonal pollen source for native honeybees in suburban ecosystems. Urban Ecosyst. 2018;21:1113–22.
186. Arcos-LeBert G, Aravena-Hidalgo T, Figueroa JA, Jaksic FM, Castro SA. Native trees provide more benefits than exotic trees when ecosystem services are weighted in Santiago. Chile Trees. 2021;35(5):1663–72.

187. Liira J, Jürjendal I. Are bees attracted by flower richness? Implications for ecosystem service-based policy. Ecol Ind. 2023;154: 110927.
188. Wu P, Axmacher JC, Song X, Zhang X, Xu H, Chen C, et al. Effects of plant diversity, vegetation composition, and habitat type on different functional trait groups of wild bees in rural Beijing. J Insect Sci. 2018;18(4):1–9.
189. Balfour NJ, Shackleton K, Arscott NA, Roll-Baldwin K, Bracuti A, Toselli G, et al. Energetic efficiency of foraging mediates bee niche partitioning. Ecology. 2021;102(4): e03285.
190. Pfeiffer VW, Zhu J, Poh K, Silbernagel J. Local influence of floral resource attributes on urban bumble bee foraging activity. Front Sustain Cities. 2023;5:1103721.
191. Hyjazie BF, Sargent RD. Floral resources predict the local bee community: implications for conservation. Biol Cons. 2022;273: 109679.
192. Fowler RE, Rotheray EL, Goulson D. Floral abundance and resource quality influence pollinator choice. Insect Conserv Div. 2016;9(6):481–94.
193. Hagler JR, Mueller S, Teuber LR, Machtley SA, Van Deynze A. Foraging range of honey bees, *Apis mellifera*, in alfalfa seed production fields. J Insect Sci. 2011;11(1):144.
194. Couvillon MJ, Riddell Pearce FC, Accleton C, Fensome KA, Quah SKL, Taylor EL, et al. Honey bee foraging distance depends on month and forage type. Apidologie. 2015;46:61–70.
195. Klein, A.-M.; Vaissière, B.E.; Cane, J.H.; Steffan-Dewenter, I.; Cunningham, S.A.; Kremen, C.; Tscharntke, T. Importance of pollinators in changing landscapes for world crops. *Proc. R. Soc. B Biol. Sci.* **2007**, *274*, 303–313.

196. Gill, R.J.; Baldock, K.C.; Brown, M.J.; Cresswell, J.E.; Dicks, L.V.; Fountain, M.T.; Garratt, M.P.; Gough, L.A.; Heard, M.S.; Holland, J.M. Protecting an ecosystem service: Approaches to understanding and mitigating threats to wild insect pollinators. In *Advances in Ecological Research*; Elsevier: Amsterdam, The Netherlands, 2016; Volume 54, pp. 135–206.
197. Potts, S.G.; Imperatriz-Fonseca, V.; Ngo, H.T.; Biesmeijer, J.C.; Breeze, T.D.; Dicks, L.V.; Garibaldi, L.A.; Hill, R.; Settele, J.; Vanbergen, A.J. *The Assessment Report on Pollinators, Pollination and Food Production: Summary for Policymakers*; Secretariat of the Intergovernmental Science-Policy Platform on Biodiversity and Ecosystem Services: Bonn, Germany, 2016.
198. Klein, A.-M.; Boreux, V.; Fornoff, F.; Mupepele, A.-C.; Pufal, G. Relevance of wild and managed bees for human well-being. *Curr. Opin. Insect Sci.* **2018**, *26*, 82–88.
199. Strange, E.E.; Ayres, M.P. *Climate Change Impacts: Encyclopedia of Life Sciences*; John Wiley & Sons: New York, NY, USA, 2010.
200. Al-Rawashdeh, Z.B.; Al-Dalaeen, J.A.; AL-Nawaiseh, M.B.; Karaje, M.R.; Al-Dalain, S.A. Impact of Climate Change on the Emergence of Plant Pathogens in Tomato Plants. *Int. J. Agric. Biosci.* **2024**, *13*, 540–546.
201. Price, M.V.; Campbell, D.R.; Waser, N.M.; Brody, A.K. Bridging the generation gap in plants: Pollination, parental fecundity, and offspring demography. *Ecology* **2008**, *89*, 1596–1604.
202. Reddy, P.; Verghese, A.; Rajan, V.V. Potential impact of climate change on honeybees (*Apis* spp.) and their pollination services. *Pest Manag. Hortic. Ecosyst.* **2012**, *18*, 121–127.

203. Giannini, T.C.; Costa, W.F.; Borges, R.C.; Miranda, L.; da Costa, C.P.W.; Saraiva, A.M.; Imperatriz Fonseca, V.L. Climate change in the Eastern Amazon: Crop-pollinator and occurrence-restricted bees are potentially more affected. *Reg. Environ. Change* **2020**, *20*, 9.

204. Mahankuda, B.; Tiwari, R. Impact of Climate Change on Honeybees and Crop Production. In *Adapting to Climate Change in Agriculture-Theories and Practices: Approaches for Adapting to Climate Change in Agriculture in India*; Sheraz Mahdi, S., Singh, R., Dhekale, B., Eds.; Springer Nature Switzerland: Cham, Switzerland, 2024; pp. 211–224.

205. Ali, M.A.; Eletmany, M.R. Climate change impacts on honeybee spread and activity: A scientific review. *Chelonian Conserv. Biol.* **2023**, *18*, 531–554.

206. Zapata-Hernández, G.; Gajardo-Rojas, M.; Calderón-Seguel, M.; Muñoz, A.A.; Yáñez, K.P.; Requier, F.; Fontúrbel, F.E.; Ormeño-Arriagada, P.I.; Arrieta, H. Advances and knowledge gaps on climate change impacts on honey bees and beekeeping: A systematic review. *Glob. Change Biol.* **2024**, *30*, e17219.

207. Potts, S.G.; Biesmeijer, J.C.; Kremen, C.; Neumann, P.; Schweiger, O.; Kunin, W.E. Global pollinator declines: Trends, impacts and drivers. *Trends Ecol. Evol.* **2010**, *25*, 345–353.

208. Pachauri, R.K.; Reisinger, A. *Climate Change 2007: Synthesis Report. Contribution of Working Groups I, II and III to the Fourth Assessment Report of the Intergovernmental Panel on Climate Change*; Ipcc: Geneva, Switzerland, 2007.

209. Nieto, A.; Roberts, S.P.M.; Kemp, J.; Rasmont, P.; Kuhlmann, M.; García Criado, M.; Biesmeijer, J.C.; Bogusch, P.; Dathe, H.H.; De la Rúa, P.; et

al. *European Red List of Bees*; Publication Office of the European Union: Luxembourg, 2014; p. 86.

210. Quaranta, M.; Cornalba, M.; Biella, P.; Comba, M.; Battistoni, A.; Rondinini, C.; Teofili, C. *Lista Rossa IUCN Delle Api Italiane Minacciate*; Comitato Italiano IUCN e Ministero dell'Ambiente e della Tutela del Territorio e del Mare: Rome, Italy, 2018; p. 68.

211. FAO. *The State of the World's Biodiversity for Food and Agriculture*; FAO: Rome, Italy, 2019; p. 572.

212. Memmott, J.; Craze, P.G.; Waser, N.M.; Price, M.V. Global warming and the disruption of plant–pollinator interactions. *Ecol. Lett.* **2007**, *10*, 710–717.

213. Hegland, S.J.; Nielsen, A.; Lázaro, A.; Bjerknes, A.L.; Totland, Ø. How does climate warming affect plant-pollinator interactions? *Ecol. Lett.* **2009**, *12*, 184–195.

214. Bartomeus, I.; Ascher, J.S.; Wagner, D.; Danforth, B.N.; Colla, S.; Kornbluth, S.; Winfree, R. Climate-associated phenological advances in bee pollinators and bee-pollinated plants. *Proc. Natl. Acad. Sci. USA* **2011**, *108*, 20645–20649.

215. Türkoğlu, N.; Şensoy, S.; Aydın, O. Effects of climate changes on phenological periods of apple, cherry and wheat in Turkey Türkiye'de iklim değişikliğinin elma, kiraz ve buğdayın fenolojik dönemlerine etkileri. *J. Hum. Sci.* **2016**, *13*, 1036–1057

216. Gérard, M.; Vanderplanck, M.; Wood, T.; Michez, D. Global warming and plant–pollinator mismatches. *Emerg. Top. Life Sci.* **2020**, *4*, 77–86

217. Landaverde, R.; Rodriguez, M.T.; Parrella, J.A. Honey Production and Climate Change: Beekeepers' Perceptions, Farm Adaptation Strategies, and Information Needs. *Insects* **2023**, *14*, 493.

218. Nemati, Z.; Ghanbari, S. Perception and adaptation strategies of beekeepers towards climate change, northwest Iran. *Int. J. Environ. Stud.* **2025**, 1–14.

219. Richardson, E.; Thompson, D. Adaptive Beekeeping Strategies to Mitigate Climate Impacts on Honey Bees in Australia. *Glob. Res. J. Eng. Sci. Technol. Manag.* **2025**, *1*, 19–27.

220. Pocol, C.B.; McDONOUGH, M. Women, Apiculture and Development: Evaluating the Impact of a Beekeeping Project on Rural Women's Livelihoods. *Bull. Univ. Agric. Sci. Vet. Med. Cluj-Napoca. Hortic.* **2015**, *72*, 487–492.

221. Prodanović, R.; Brkić, I.; Soleša, K.; Ljubojević Pelić, D.; Pelić, M.; Bursić, V.; Vapa Tankosić, J. Beekeeping as a tool for sustainable rural development. *J. Agron. Technol. Eng. Manag.* **2024**, *7*, 1054–1066.

222. Herab, A.; Al-Ghamdi, A.; Alzahrani, K.; Elhindi, K.M.; Muddassir, M.; Kassem, H.S. A framework for quantifying the strength of partnerships between agricultural cooperatives and development actors: A case study in saudi arabia. *Int. J. Environ. Res. Public Health* **2022**, *20*, 364.

223. Lecoutere, E.; Spielman, D.J.; Van Campenhout, B. Empowering women through targeting information or role models: Evidence from an experiment in agricultural extension in Uganda. *World Dev.* **2023**, *167*, 106240

224. Mishra, G.; Suryavanshi, A.; Pathak, D.K. Women Empowerment through Agriculture Extension Services. In *Advanced Trends in Agricultural Extension*; Integrated Publications: Rohini, India, 2024; pp. 41–52.

225. Quisumbing, A.; Cole, S.; Elias, M.; Faas, S.; Galiè, A.; Malapit, H.; Meinzen-Dick, R.; Myers, E.; Seymour, G.; Twyman, J. Measuring Women's Empowerment in Agriculture: Innovations and evidence. *Glob. Food Secur.* **2023**, *38*, 100707.

226. Costa, V.; Piedrahita, N.; Mane, E.; Davis, B.; Slavchevska, V.; Gurbuzer, Y.L. *Women's Employment in Agrifood Systems. Background Paper for the Status of Women in Agrifood Systems*; FAO: Rome, Italy, 2023; Available online: **https://openknowledge.fao.org/server/api/core/bitstreams/e7689bf7-00f0-465b-ad03-e0c56ffb14b1/content** (accessed on 20 April 2025).
227. Sabourin, A.; Jones, A. *More Than Targets: How the EU Promotes Democracy, Human Rights and Gender Equality Through Global Europe and Beyond*; ECDPM Discussion Paper 337; ECDPM: Maastricht, The Netherlands, 2023; Available online: **https://ecdpm.org/application/files/3316/7845/4683/How-EU-Promotes-Democracy-Human-Rights-Gender-Equality-Global-Europe-Beyond-ECDPM-Discussion-Paper-337-2023.pdf** (accessed on 20 April 2025).
228. Salzinger, M.; Desmidt, S. *Gender Equality in EU External Action: The Gender Action Plan and the Women, Peace and Security Agenda*; ECDPM: Maastricht, The Netherlands, 2023.
229. FAO. FAO Regional Office for Near East and North Africa. Available online: **https://www.fao.org/neareast/news/stories/details/towards-a-sustainable-future--empowering-rural-women-in-saudi-arabia-through-sustainable-rural-agriculture/en#:~:text=The%20government%20in%20Saudi%20Arabia,rural%20women%20to%20accelerate%20progress.&text=With%20technical%20support%20from%20FAO,Agriculture%20Development%20(SRAD)%20programme** (accessed on 12 February 2025).

230. GOV.SA. Reef National Foundation Launches Initiative to Empower Rural Women across Saudi Arabia. Available online: **https://my.gov.sa/en/news/16566** (accessed on 18 April 2025).

231. Agency, S.P. Saudi Reef Program Backs Rural Development with SAR 2.4 Billion Since Launch. Available online: **https://www.spa.gov.sa/en/N2242136.** (accessed on 18 April 2025).

232. Ulllah, M.; Alsanhani, A.; Aldawdahi, N. Farmer's perception of climate change: An assessment from medina region, Saudi Arabia. *Agrobiol. Rec.* **2024**, *18*, 12–17.

233. Al-Ghamdi, A.; Nuru, A. Beekeeping in the Kingdom of Saudi Arabia opportunities and challenges. *Bee World* **2013**, *90*, 54–57.

234. Vinci, G.; Rapa, M.; Roscioli, F. Sustainable development in rural areas of Mexico through beekeeping. *Int. J. Sci. Eng. Invent.* **2018**, *4*, 1–7

235. Nat Schouten, C.; John Lloyd, D. Considerations and factors influencing the success of beekeeping programs in developing countries. *Bee World* **2019**, *96*, 75–80.

236. Alarify, I.A.; Rawwash, A.A.E.M.M.; Mohmmed, S.E.R. Comparing Saudi Honey with Honey in the Saudi Market-A Statistical Study. *Spec. Ugdym.* **2022**, *1*, 43.

237. CDSI. *Central Department of Statistics & Information, Saudi Arabia*; Institute for Health Metrics and Evaluation: Seattle, WA, USA, 2010.

238. FAO. *Beekeeping and Honey Production Sector Review and Situation Analysis in the Kingdom of Saudi Arabia*; Ministry of Environment, Water and Agriculture Saudi Arabia: Riyadh, Saudi Arabia, 2021. Available online: **https://www.mewa.gov.sa/reefea/attachments/%D9%82%D8%B7%D8%A7%D8%B9%20%D8%A7%D9%84%D8%B9%D8%B3%D**

9%84/BEE-2021-Honey%20Sector%20Review-TC.pdf (accessed on 1 April 2025).

239. Hamza, R.M.A.; Aslam, H.; Aleem, K.; Bukhari, M.A.; Nadeem, J.; Farooq, H.; Akbar, M.; Manzoor, M.; Ashraf, H. The genetics of insecticide resistance in relation to climate change. *Agrobiol. Rec.* **2023**, *12*, 92–98.
240. World Bank. How Rural Development and Agriculture Can Support Social Inclusion in Developing Countries—A Sub-Saharan Approach. Jemilehin funmilayoelsy Buligwa Asenanewton Muchiti Mayulammatshepo Seoka. 14 November 2022. Available online: **https://blogs.worldbank.org/en/youth-transforming-africa/how-rural-development-and-agriculture-can-support-social-inclusion** (accessed on 9 January 2025).
241. UN Women. Rural Women, Official Web of Sit, Data Inter 15/2/2024 Rural Women | What We Do: Economic Empowerment | UN Women—Headquarters. 2024. Available online: **https://www.unwomen.org/en/what-we-do/economic-empowerment/rural-women** (accessed on 15 April 2025).
242. Al-Khalayleh, M.A.H.A. The Role of Municipalities in Empowering Women in Rural Communities. *Int. J. Humanit. Soc. Sci.* **2022**, *39*, 83–94.
243. United Nations General Assembly Economic and Social Council. Joint Action Plan of the United Nations Entity for Gender Equality and the Empowerment of Women and the Office of the United Nations High Commissioner for Human Rights. 2012. Available online: **https://documents.un.org/doc/undoc/gen/n11/631/97/pdf/n1163197.pdf** (accessed on 20 April 2025).

244. Newspaper, Alriyadh. ―Saudi Reef": SAR 140 million to Support the Honey Sector 18 April 2024. Available online: **https://www.alriyadh.com/2070582** (accessed on 22 April 2025).
245. Al-Turaif, G.A.; Al-Rashidi, A. Planning Indicators to Empower Saudi Women in the Fields of Tourism Investment in Light of Vision 2030. *Al-Qura Univ. J.* **2019**, *12*. Available online: **https://research.ebsco.com/linkprocessor/plink?id=b1f25d19-acb8-3fc6-9d5e-c5992902ee66** (accessed on 13 March 2025).
246. Food and Agriculture Organization of the United Nations. Practical Manual on Organic Beekeeping; BEE/051/2022/5; Riyadh, Kingdom of Saudi Arabia. 2022. Available online: **https://www.mewa.gov.sa/reefea/attachments/%D9%82%D8%B7%D8%A7%D8%B9%20%D8%A7%D9%84%D8%B9%D8%B3%D9%84/BEE-2022-Manual%20on%20organic%20beekeeping.pdf** (accessed on 13 March 2025).
247. Patel, V.; Pauli, N.; Biggs, E.; Barbour, L.; Boruff, B. Why bees are critical for achieving sustainable development. *Ambio* **2021**, *50*, 49–59.
248. Emirate of Hail Region About the Region, the Website of the Emirate of Hail Region, Kingdom of Saudi Arabia, Retrieved Date. 2024. Available online: **https://hail.gov.sa/en/node/1109** (accessed on 18 September 2024).
249. Hereher, M.E.; Al-Shammari, A.M.; Abd Allah, S.E. Land cover classification of Hail—Saudi Arabia using remote sensing. *Int. J. Geosci.* **2012**, *3*, 349.
250. Saudi Ministry of Investment. An Overview of the Hail Region, Invest in Saudi Arabia, Saudi Arabia. Available

online: **https://www.investsaudi.sa/ar/meetTheKingdom/province/hail** (a ccessed on 18 September 2024).

251. Thompson, S.K. *Sampling*; John Wiley & Sons: New York, NY, USA, 2012; Volume 755.
252. Bocianowski, J.; Wrońska-Pilarek, D.; Krysztofiak-Kaniewska, A.; Matusiak, K.; Wiatrowska, B. Comparison of Pearson's and Spearman's correlation coefficients values for selected traits of *Pinus sylvestris* L. *Res. Sq.* **2024**, 1–20.
253. Strong, M.M. *The Effect of Farmer Chacteristics on Entrepreneurial Behavior of Beekeepers in Kibwezi West Sub County, Makueni County*; Strathmore University: Nairobi, Kenya, 2021.
254. Honeycutt, C.M. *Profitable, Alternative Income Generation and Improved Quality of Life Among Global Beekeepers*; East Tennessee State University: Johnson, TN, USA, 2023.
255. Shah, J.A.; Asmuni, A.; Ismail, A. Roles of extension agents towards agricultural practice in Malaysia. *Int. J. Adv. Sci. Eng. Inf. Technol.* **2013**, *3*, 59–63.
256. Muddassir, M.; Alotaibi, B.A. Farmers' awareness of improved maize cropping practices in Punjab, Pakistan. *Int. J. Agric. Biosci.* **2023**, *12*, 8–17.
257. Muddassir, M.; Al Shenaifi, M.S.; Kassem, H.S.; Alotaibi, B.A. Adoption of improved maize production technologies in Punjab Province, Pakistan. *J. Agric. Ext.* **2020**, *24*, 1–11.
258. Benge, M.; Amy, V. Challenges and barriers to conducting commercial beekeeping education programs in Florida. *Adv. Agric. Dev.* **2023**, *4*, 49–57.

259. Gikunda, R.M.; Ooga, D.M.; Okiamba, I.N.; Anyuor, S. Cultural barriers towards women and youth entry to apiculture production in Maara Sub-County, Kenya. *Adv. Agric. Dev.* **2021**, *2*, 73–85.

260. Anaeto, F.C.; Asiabaka, C.; Nnadi, F.; Ajaero, J.; Aja, O.; Ugwoke, F.; Ukpongson, M.; Onweagba, A. The role of extension officers and extension services in the development of agriculture in Nigeria. *J. Agric. Res.* **2012**, *1*, 180–185.

261. Musa, M.; Ismail, M.M.; Ismail, W.I. Effectiveness of extension agent services in influencing the adoption of modern hive in sustainable stingless beekeeping. *J. Sustain. Sci. Manag.* **2019**, *14*, 14–24.

262. Ismail, M.M.; Ismail, W.I.W. Development of stingless beekeeping projects in Malaysia. *E3S Web Conf.* **2018**, *52*, 00028

263. Mburu, P.D.M.; Affognon, H.; Irungu, P.; Mburu, J.; Raina, S. Gender roles and constraints in beekeeping: A case from Kitui County, Kenya. *Bee World* **2017**, *94*, 54–59.

264. Univerity, K.K. Beekeeping and Bee Products" Training Program. 2025. Available online: **https://www.kku.edu.sa/en/news/100767** (accessed on 8 January 2025).

265. Feketéné Ferenczi, A.; Szűcs, I.; Bauerné Gáthy, A. ―What's Good for the Bees Will Be Good for Us!"—A Qualitative Study of the Factors Influencing Beekeeping Activity. *Agriculture* **2024**, *14*, 890

266. Azeez, K.K. Analysis of information needs of beekeepers in Oyo State, Nigeria. *J. Agric. Res. Dev. Ext. Technol.* **2022**, *4*, 18–36.

267. Amulen, D.R.; D'Haese, M.; Ahikiriza, E.; Agea, J.G.; Jacobs, F.J.; de Graaf, D.C.; Smagghe, G.; Cross, P. The buzz about bees and poverty

alleviation: Identifying drivers and barriers of beekeeping in sub-Saharan Africa. *PLoS ONE* **2017**, *12*, e0172820.

268. Mujuni, A.; Natukunda, K.; Kugonza, D.R. Factors affecting the adoption of beekeeping and associated technologies in Bushenyi District, Western Uganda. *Development* **2012**, *24*, 1–19.

269. Mulatu, A.; Marisennayya, S.; Bojago, E. Adoption of Modern Hive Beekeeping Technology: The Case of Kacha-Birra Woreda, Kembata Tembaro Zone, Southern Ethiopia. *Adv. Agric.* **2021**, *2021*, 4714020.

270. Maderson, S. Co-producing agricultural policy with beekeepers: Obstacles and opportunities. *Land Use Policy* **2023**, *128*, 106603.

271. Benin, S.; Nkonya, E.; Okecho, G.; Pender, J.; Nahdy, S.; Mugarura, S. *Assessing the Impact of the National Agricultural Advisory Services (NAADS) in the Uganda Rural Livelihoods*; International Food Policy Research Institute: Washington, DC, USA, 2007.

272. Ramirez, A. The influence of social networks on agricultural technology adoption. *Procedia-Soc. Behav. Sci.* **2013**, *79*, 101–116.

273. Girma, J.; Gardebroek, C. The impact of contracts on organic honey producers‘ incomes in southwestern Ethiopia. *For. Policy Econ.* **2015**, *50*, 259–268.

274. Kassem, H.S.; Alotaibi, B.A.; Muddassir, M.; Herab, A. Factors influencing farmers‘ satisfaction with the quality of agricultural extension services. *Eval. Program Plan.* **2021**, *85*, 101912.

275. Dabiah, A.T. Rural community‘s participation rate and perceived benefits in a rural area in Saudi Arabia. *Agrobiol. Rec.* **2023**, *12*, 1–8.

276. Verbeke, W.; Diallo, M.A.; van Dooremalen, C.; Schoonman, M.; Williams, J.H.; Van Espen, M.; D‘Haese, M.; de Graaf, D.C. European beekeepers‘

interest in digital monitoring technology adoption for improved beehive management. *Comput. Electron. Agric.* **2024**, *227*, 109556.

277. Motti, R., Paura, B., Cozzolino, A. & Falco, B. Edible flowers used in some countries of the Mediterranean Basin: An ethnobotanical overview. *Plants* **11**(23), 3272 (2022).

278. Bokelmann, J. M. Fennel (Foeniculum vulgare). In *Medicinal Herbs in Primary Care* (ed. Bokelmann, J. M.) 335–340 (Elsevier, 2022).

279. Soumanou, M. M. & Adjou, E. S. Sweet Fennel (Ocimum gratissimum) Oils. In *Essential Oils in Food Preservation, Flavor and Safety* (ed. Preedy, V. R.) 765–773 (Academic Press, 2016).

280. Singh, Y. V. *et al.* Effect of integrated nutrient management on growth parameters of fennel (Foeniculam valgare Mill) in inceptisol. *Int. J. Curr. Microbiol. Appl. Sci.* **7**(03), 1067–1072. https://doi.org/10.20546/ijcmas.2018.703.127 (2018).

281. Mehra, N., Tamta, G. & Nand, V. A review on nutritional value, phytochemical and pharmacological attributes of Foeniculum vulgare Mill. *J. Pharmacognosy Phytochem.* **10**(2), 1255–1263. https://doi.org/10.22271/phyto.2021.v10.i2q.13983 (2021).

282. Agarwal, D., Saxena, S. N., Sharma, L. K. & Lal, G. Prevalence of essential and fatty oil constituents in fennel (Foeniculum vulgare Mill) genotypes grown in semi arid regions of India. *J. Essential Oil Bearing Plants* **21**(1), 40–51. https://doi.org/10.1080/0972060x.2018.1433072 (2018).

283. Verma, A. K., Sharma, S., Kakani, R. K., Meena, R. D. & Choudhary, S. Gamma radiation effects seed germination, plant growth and yield attributing characters of fennel (Foeniculum vulgare Mill). *Int. J. Curr.*

Microbiol. Appl. Sci. **6**(5), 2448–2458. https://doi.org/10.20546/ijcmas.2017.605.274 (2017).

284. Ebeed, N. M., Abdou, H. S., Booles, H. F., Salah, S., & Ahmed, E. S. (2010). *Antimutagenic and Chemoprevention Potentialities of Sweet Fennel (Foeniculum vulgare Mill.) Hot Water Crude Extract*.

285. Bharti, V. *et al.* Diversity, abundance and pollination efficiency of insect pollinators of fennel (Foeniculum vulgare Miller) and effect of abiotic factors on insect pollinator activity. *J. Appl. Nat. Sci.* **7**(2), 786–793 (2015).

286. Shilpa, P., Sowmya, K., Srikanth, C. & Kuberappa, G. Pollinator diversity and foraging activity on fennel, Foeniculum vulgare Mill and African marigold, Tagetus minuta L. *Pest Manag. Horticult. Ecosyst.* **20**(2), 236–239 (2014).

287. Wu, Y. & Yuan, H. Research review of insect pollination mechanism. *J. Jilin Agric. Univ.* **27**(6), 608–613 (2005).

288. Telci, I., Demirtas, I. & Sahin, A. Variation in plant properties and essential oil composition of sweet fennel (Foeniculum vulgare Mill) fruits during stages of maturity. *Ind. Crops Products.* **30**(1), 126–130. https://doi.org/10.1016/j.indcrop.2009.02.010 (2009).

289. Harris, B. A., Braman, S. K. & Pennisi, S. V. Influence of plant taxa on pollinator, butterfly, and beneficial insect visitation. *HortScience* **51**(8), 1016–1019. https://doi.org/10.21273/Hortsci.51.8.1016 (2016).

290. Deora, T., Ahmed, M. A., Daniel, T. L. & Brunton, B. W. Tactile active sensing in an insect plant pollinator. *J. Exp. Biol.* **224**(Pt 4), jeb239442. https://doi.org/10.1242/jeb.239442 (2021).

291. Etter, K. J. *et al.* Interspecific pollen transport between non-native fennel and an island endemic buckwheat: Assessment of the magnet effect. *Biol.*

Invasions **24**(1), 139–155. https://doi.org/10.1007/s10530-021-02626-0 (2022).

292. Nicolson, S. W. Bee food: The chemistry and nutritional value of nectar, pollen and mixtures of the two. *Afr. Zool.* **46**(2), 197–204. https://doi.org/10.3377/004.046.0201 (2011).
293. Ricigliano, V. A. *et al.* The impact of pollen consumption on honey bee (*Apis mellifera*) digestive physiology and carbohydrate metabolism. *Arch. Insect. Biochem. Physiol.* **96**(2), e21406. https://doi.org/10.1002/arch.21406 (2017).
294. Palmer-Young, E. C. *et al.* Nectar and pollen phytochemicals stimulate honey bee (Hymenoptera: Apidae) immunity to viral infection. *J. Econ. Entomol.* **110**(5), 1959–1972. https://doi.org/10.1093/jee/tox193 (2017).
295. Tsuruda, J. M. & Page, R. E. The effects of young brood on the foraging behavior of two strains of honey bees (*Apis mellifera*). *Behav. Ecol. Sociobiol.* **64**(2), 161–167. https://doi.org/10.1007/s00265-009-0833-3 (2009).
296. Wilkaniec, Z., Giejdasz, K. & Proszynski, G. Effect of pollination of onion seeds under isolation by the red mason bee (Osmia rufa L) (Apoidea, Megachilidae) on the setting and quality of obtained seeds. *J. Apicult. Sci.* **48**(2), 35–41 (2004).
297. Rasekh, A. M., Sadeghi, N. H., & Hosseini, M. (2013). Effects of insect pollinators on onion seed production quality and quantity.
298. Arroyo-Correa, B., Beattie, C. & Vallejo-Marin, M. Bee and floral traits affect the characteristics of the vibrations experienced by flowers during buzz pollination. *J. Exp. Biol.* https://doi.org/10.1242/jeb.198176 (2019).

299. Brito, V. L. G., Nunes, C. E. P., Resende, C. R., Montealegre-Zapata, F. & Vallejo-Marin, M. Biomechanical properties of a buzz-pollinated flower. *R. Soc. Open Sci.* **7**(9), 201010. https://doi.org/10.1098/rsos.201010 (2020).

300. Alqarni, A. S. Honey bee foraging, nectar secretion, and honey potential of Wild Jujube Trees, Ziziphus nummularia. *Neotrop. Entomol.* **44**(3), 232–241. https://doi.org/10.1007/s13744-015-0279-4 (2015).

301. El-Shakaa, S. M. (1977). *Studies on pollen grains collected by the honey bees in Giza region.* (M.Sc. M.Sc.). Cairo Univ,

302. Layek, U., Das, A. & Karmakar, P. Supplemental stingless bee pollination in fennel (Foeniculum vulgare Mill): An assessment of impacts on native pollinators and crop yield. *Front. Sustain. Food Syst.* https://doi.org/10.3389/fsufs.2022.820264 (2022).

303. Kumar, M. & Singh, R. Role of honey bees in the pollination of fennel (Foeniculum vulgare L). *J. Pharmacognosy Phytochem.* **6**(1), 214–218 (2017).

304. Carisio, L. *et al.* Estimates of nectar productivity through a simulation approach differ from the nectar produced in 24 h. *Funct. Ecol.* **36**(12), 3234–3247. https://doi.org/10.1111/1365-2435.14210 (2022).

305. Waser, N. M. Interspecific pollen transfer and competition between co-occurring plant species. *Oecologia.* **36**(2), 223–236. https://doi.org/10.1007/BF00349811 (1978).

306. Kearns CA, Inouye DW. Pollinators, flowering plants, and conservation biology. Bioscience. 1997;47:297-307

307. Ollerton J et al. How many flowering plants are pollinated by animals? Oikos. 2011;120:321-326

308. Cox PA, Elmqvist T. Pollinator extinction in the Pacific Islands. Conservation Biology. 2000;14:1237-1239

309. Klein A-M, Vaissière B, Cane JH, Steffan-Dewenter I, Cunningham SA, Kremen C, et al. Importance of pollinators in changing landscapes for world crops. Proceedings of the Royal Society B. 2007;274:303-313

310. Gallai N, Salles JM, Settele J, Vaissiere BE. Economic valuation of the vulnerability of world agriculture confronted with pollinator decline. Ecological Economics. 2009;68:810-821

311. Barfield A, Bergstrom J, Ferreira S. An Economic Valuation of Pollination Services in Georgia. Birmingham: Southern Agricultural Economics Association; 2012

312. Jian D, Chen WF. Economic value of insect pollination for fruits and vegetables in China. Acta Entomologica Sinica. 2011;54(4):443-450

313. Aizen MA, Morales CL, Morales JM. Invasive mutualists erode native pollination Webs. PLoS Biology. 2008;6(2):e31. DOI: 10.1371/journal.pbio.0060031

314. Aizen MA, Harder LD. The Global stock of domesticated honey bees is growing slower than agricultural demand for pollination. Current Biology. 2009;19:1-4

315. Kevan PG, Baker HG. Insects as flower visitors and pollinators. Annual Review of Entomology. 1983;28:407-453

316. Losey JE, Vaughan M. The economic value of ecological services provided by insects. Bioscience. 2006;56:311-323

317. Robbins CS, Sauer JR, Greenberg RS, Droege S. Population declines in North American birds that migrate to the neotropics. Proceedings of the National Academy of Sciences USA. 1989;86(19):7658-7662

318. Free JB. Insect Pollination of Crops. London: Academic Press; 1970. pp. 544
319. McGregor SE. Insect pollination of cultivated crop plants. USDA Handbook 496. Washington: US Department of Agriculture, Agricultural Research Services; 1976. pp. 411
320. Pesson P, Louveaux J, editors. Pollination et Productions Vegetables. Paris: INRA; 1984
321. Free JB. Insect Pollination of Crops. 2nd ed. London, UK: Academic Press; 1993
322. Cane JH. Alkali Bee. In: Shimanuki H, Flottum K, Harman A, editors. ABC & XYZ of Bee Culture. 41st ed. A.I. Root Company: Medina, OH; 2007. pp. 11-12
323. Elfattah MSA. Sustainable use of leafcutting bee hives for alfalfa pollination. Munis Entomology & Zoology. 2010;5(2):807-811
324. Mader E, Spivak M, Evans E. Managing Alternative Pollinators: A Handbook for Beekeepers, Growers, and Conservationists. Beltsville: Sustainable Agriculture Network; 2010. pp. 167
325. Richards KW. Alfalfa leafcutter bee management in Western Canada. Ottawa: Agriculture Canada; 1984. pp. 51
326. Torchio PF. Use of non-honey bee species as pollinators of crops. Proceedings of the Entomological Society of Ontario. 1987;118:111-124
327. Plowright RC, Laverty TM. Bumble bees and crop pollination in Ontario. Proceedings of the Entomological Society of Ontario. 1987;118:155-160
328. Pellmyr O. Yuccas, yucca moths, and coevolution: A review. Annals of the Missouri Botanical Garden. 2003;90:35-55

329. Kevan PG. Pollination by animals and angiosperm biosystematics. In: Grant WF, editor. Plant Biosystematics. Toronto: Academic Press; 1984. pp. 271-292

330. Tepedino VJ. The pollination efficiency of the squash bee (Peponapispruinosa) and the honey bee (Apismellifera) on summer squash (Cucurbitapepo). Journal of the Kansas Entomological Society. 1981;54:359-377

331. Parker FD, Frohlich DR. Studies on management of the sunflower leafcutter bee Eumegachilepugnata (Say) (Hymentoptera: Megachilidae). Journal of Apicultural Research. 1985;24:125-131

332. Sanchez D. Ground nesting bee best for blueberries. Agricultural Research. 1989;37(6):16-17

333. Bosch J, Kemp W. How to Manage the Blue Orchard Bee as an Orchard Pollinator. Beltsville: Sustainable Agriculture Network; 2001. 88 pp. Available from: www.sare.org/publications/bee/blue_orchard_bee.pdf

334. Torchio PF. Diversification of pollination strategies for U.S. crops. Environmental Entomology. 1990;19(6):1649-1656

335. Torchio PF. Osmia ribifloris a native bee species developed as a commercially managed pollinator of highbush blueberry. Journal of the Kansas Entomological Society. 1990;63:427-436

336. Syed RA. Studies on oil palm pollination by insects. Bulletin of Entomological Research. 1978;69:213-224

337. Pouvreau A. Pollinisation du cacotier, du papayer, ettucafeier. In: Pesson P, Louveaux J, editors. Pollination et Productions Vegetales. Paris: INRA; 1984. pp. 195-208

338. Kral R. A revision of Asimina and Deeringothamus (Annonaceae). Brittonia. 1960;12:233-278
339. Norman EM, Clayton D. Reproductive biology of two Florida pawpaws; Asmina ovata and A pygmaea. Bulletin of the Torrey Botanical Club. 1986;113:16-22
340. Grant V. The pollination of Calycanthusoccidentalis. American Journal of Botany. 1950;37:294-297
341. Dilley JD. The radiation of Calochortus: Generalist flowers moving through a mosaic of potential pollinators. Oikos. 2000;89:209-222
342. Grant KA, Grant V. Flower Pollination in the Phlox Family. New York: Columbia University Press; 1965
343. Bloch D, Werdenberg N, Erhardt A. Pollination crisis in the butterfly pollinated, wild carnation Dianthus carthusianorum? New Phytologist. 2006;169:699-706
344. Ehrlich PR, Hanski I, editors. On the Wings of Checkerspots: A Model System for Population Biology. New York: Oxford University Press; 2004
345. Halpern S. Four Wings and a Prayer: Caught in the Mystery of the Monarch Butterfly. New York: Pantheon; 2001
346. Boggs CL, Smiley JT, Gilbert LE. Patterns of pollen exploitation by Heliconius butterflies. Oecologia. 1981;48:284-289
347. Estrada C, Jiggins CD. Patterns of pollen feeding and habitat preference among Heliconius species. Ecological Entomology. 2002;27:448-456
348. Pellmyr O, Thompson JN, Brown JM, Harrison RG. Evolution of pollination and mutualism in the yucca moth lineage. American Naturalist. 1996;148:827-847

349. Gregory D. Hawkmoth pollination in the genus Oenothera. Aliso. 1963-1964;5:357-419
350. Lin S, Bernardello G. Flower structure and reproductive biology in Aspidospermaquebracho-blanco (Apocynaceae), a tree pollinated by deceit. International Journal of Plant Sciences. 1999;160:869-878
351. Little KJ, Dieringer G, Romano M. Pollination ecology, genetic diversity and selection on nectar spur length in Platantheralacera (Orchidaceae). Plant Species Biology. 2005;20:183-190
352. Raguso RA, Henzel C, Buchmann SL, Nabhan GP. Trumpet flowers of the Sonoran Desert: Floral biology of Peniocereuscacti and Sacred Datura. International Journal of Plant Sciences. 2003;164:877-892
353. Ananthakrishnan TN. The role of thrips in pollination. Current Science (Bangalore). 1993;65(3):262-264
354. Kirk WDJ. Thrips and pollination biology. In: Ananthakrishnan TN, Raman A, editors. Dynamics of Insect-Plant Interaction. New Delhi: Oxford and IBH; 1988. pp. 129-135. [Chapter 10]
355. Lewis T. Thrips: Their Biology, Ecology and Economic Importance. London and New York: Academic Press; 1973
356. Terry I. Host selection, communication, and reproductive behaviour. In: Lewis T, editor. Thrips as Crop Pests. London: CAB International; 1997. pp. 64-118
357. Terry I. Thrips and weevils as dual, specialist pollinators of the Australian cycad Macrozamiacommunis (Zamiaceae). International Journal of Plant Sciences. 2001;162:1293-1305

358. Albrecht M, Schmid B, Hautier Y, Mueller CB. Diverse pollinator communities enhance plant reproductive success. Proceedings of the Royal Society B-Biological Sciences. 2012;279:4845-4852

359. Davies TJ, Barraclough TG, Chase MW, Soltis PS, Soltis DE, Savolainen V. Darwin's abominable mystery: insights from a supertree of the angiosperms. Proceedings of the National Academy of Sciences, USA. 2004;101(7):1904-1909

360. Labandeira CC, Dilcher DL, Davis DR, Wagner DL. Ninety-seven million years of angiosperm-insect association: Paleobiological insights into the meaning of coevolution. Proceedings of the National Academy of Sciences USA. 1994;91:12278-12282

361. Simpson BB, Neff JL. Evolution and diversity of floral rewards. In: Jones CE, Little RJ, editors. Handbook of Experimental Pollination Biology. New York: Van Nostrand Reinhold Co.; 1983. pp. 142-159

362. Bond WJ. Do mutualisms matter? Assessing the impact of pollinator and disperser disruption on plant extinction. Philosophical Transactions of the Royal Society of London Series B—Biological Sciences. 1994;344:83-90

363. Ollerton J. Pollinator diversity: Distribution, ecological function, and conservation. In: Futuyma DJ, editor. Annual Review of Ecology, Evolution, and Systematics. Vol. 48. 2017. pp. 353-376

364. Wilcock C, Neiland R. Pollination failure in plants: Why it happens and when it matters. Trends in Plant Science. 2002;7:270-277

365. Ratto F, Simmons BI, Spake R, Zamora-Gutierrez V, MacDonald MA, Merriman JC, et al. Global importance of vertebrate pollinators for plant reproductive success: A meta-analysis. Frontiers in Ecology and the Environment. 2018;16(2):82-90

366. Fort H. Quantitative predictions of pollinators' abundances from qualitative data on their interactions with plants and evidences of emergent neutrality. Oikos. 2014;123:1469-1478
367. Kaiser-Bunbury CN, Muff S, Memmott J, Muller CB, Caflisch A. The robustness of pollination networks to the loss of species and interactions: A quantitative approach incorporating pollinator behaviour. Ecology Letters. 2010;13:442-452
368. Memmott J, Craze PG, Waser NM, Price MV. Global warming and the disruption of plant-pollinator interactions. Ecology Letters. 2007;10:710-717
369. Roulston TH, Cane JH. Pollen nutritional content and digestibility for animals. Plant Systematics and Evolution. 2000;222:187-209
370. Muller A, Kuhlmann M. Pollen hosts of western palaearctic bees of the genus Colletes (Hymenoptera: Colletidae): The Asteraceae paradox. Biological Journal of the Linnean Society. 2008;95:719-733
371. Viana BF, Boscolo D, Neto EM, Lopes LE, Lopes AV, Ferreira PA, et al. How well do we understand landscape effects on pollinators and pollination services? Journal of Pollination Ecology. 2012;7:31-41
372. Nicholls E, Hempel de Ibarra N. Assessment of pollen rewards by foraging bees. Functional Ecology. 2017;31:76-87
373. Ruedenauer FA, Spaethe J, Leonhardt SD. How to know which food is good for you: Bumblebees use taste to discriminate between different concentrations of food differing in nutrient content. Journal of Experimental Biology. 2015;218:2233-2240
374. Armbruster WS. The specialization continuum in pollination systems: Diversity of concepts and implications for ecology, evolution and conservation. Functional Ecology. 2017;31:88-100

375. Cameron SA, Lozier JD, Strange JP, Koch JB, Cordes N, Solter LF, et al. Patterns of widespread decline in North American bumble bees. Proceedings of the National Academy of Sciences of the United States of America. 2011;108:662-667

376. Cameron SA et al. Patterns of widespread decline in North America bumble bee. Proceedings of the National Academy of Sciences of the United States of America. 2011;108:662-667

377. Goulson D, Lye GC, Darvill B. Decline and conservation of bumble bees. Annual Review of Entomology. 2009;53:191-210

378. Kerr JT, Pindar A, Galpern P, Potts SG, Roberts SM, Rasmont P, et al. Climate change impacts on bumblebees converge across continents. Science. 2015;349:177-180

379. Potts SG, Biesmeijer JC, Kremen C, Neumann P, Schweiger O, Kunin WE. Global pollinator declines: Trends, impacts and drivers. Trends in Ecology and Evolution. 2010;25:345-353

380. Aizen MA, Harder LD. Expanding the limits of the pollen limitation concept: Effects of pollen quantity and quality. Ecology. 2007;88:271-281

381. Morse RA, Flottum K. Honey Bee Pests, Predators, and Diseases. 3rd ed. Medina, Ohio: The A. I. Root Company; 1997. 718 pp

382. Barrett DP. History of American foulbrood in Michigan: Period 1927-1954. Gleanings in Bee Culture. 1955;83:460-461

383. Evans JD. Diverse origins of tetracycline resistance in the honey bee bacterial pathogen Paenibacillus larvae. Journal of Invertebrate Pathology. 2003;83:46-50

384. Gary NE Jr, Page RE, Morse RA, Henderson CE, Nasr ME, Lorenzen K. Comparative resistance of honey bees (Apismellifera L.) from Great Britain

and United States to infestation by tracheal mites (Acarapiswoodi). American Bee Journal. 1990;130:667-669

385. Kevan PG, Greco CF, Belaoussoff S. Log-normality of biodiversity and abundance in diagnosis and measuring of ecosystemic health: Pesticide stress on pollinators on blueberry heaths. Journal of Applied Ecology. 1997;34:1122-1136

386. Nasr ME, Otis GW, Scott-Dupree CD. Resistance to Acarapiswoodi by honey bees (Hymenoptera: Apidae): Divergent selection and evaluation of selection progress. Journal of Economic Entomology. 2001;94(2):332-338

387. Oldroyd BP, Wongsiri S. Asian Honey Bees: Biology, Conservation, and Human Interactions. Cambridge, Mass: Harvard University Press; 2006

388. Decourtye A, Devillers J, Genecque E, Le Menach K, Budzinski H, Cluzeau S, et al. Comparative sublethal toxicity of nine pesticides on olfactory performances of the honeybee Apismellifera. Pesticide Biochemistry and Physiology. 2004;78:83-92

389. Desneaux N, Decourtye A, Delpuech J. The sublethal effects of pesticides on beneficial arthropods. Annual Review of Entomology. 2007;52:81-106

390. Johansen CA, Mayer DF. Pollinator Protection. A Bee and Pesticide Handbook. Cheshire, Conn: Wicwas Press; 1990

391. Thompson HM. Behavioural effects of pesticides use in bees—Their potential for use in risk assessment. Ecotoxicology. 2003;12:317-330

392. Losey JE, Rayor LS, Carter ME. Transgenic pollen harms monarch larvae. Nature. 1999;399:214

393. Malone LA, Pham-Delègue MH. Effects of transgene products on honey bees (Apismellifera) and bumblebees (Bombussp.). Apidologie. 2001;32:287-304

394. O'Callaghan M, Glare TR, Burgess EPJ, Malone LA. Effects on plants genetically modified for insect resistance on non-target organisms. Annual Review of Entomology. 2005;50:271-292
395. National Research Council (NRC). Status of Pollinators in North America. Washington, DC: The National Academies Press; 2007
396. Kevan PG. Pollination: Plinth, pedestal, and pillar for terrestrial productivity. The why, how, and where of pollination protection, conservation, and promotion. In: Stubbs CS, Drummond FA, editors. Bees and Crop Pollination—Crisis, Crossroads, Conservation, Proceedings, Entomological Society of America. Lanham, Md.: Thomas Say Publications in Entomology; 2001. pp. 7-68
397. Kearns CA, Inouye DW, Waser NM. Endangered mutualisms: The conservation of plant-pollinator interactions. Annual Review of Ecology and Systematics. 1998;29:83-112
398. Gess FW, Gess SK. Effects of increasing land utilization on species representation and diversity of aculeate wasps and bees in the semi-arid areas of southern Africa. In: LaSalle J, Gauld ID, editors. Hymenoptera and Biodiversity. Wallingford: CAB International; 1993. pp. 83-113
399. Goulson D. Effects of introduced bees on native ecosystems. Annual Review of Ecology and Systematics. 2003;34:1-26
400. Kevan PG. Pollinators as bioindicators of the state of the environment: Species, activity and diversity. Agriculture, Ecosystems and Environment. 1999;74:373-393
401. Vinson SB, Frankie GW, Barthell J. Threats to the diversity of solitary bees in a neotropical dry forest in Central America. In: LaSalle J, Gauld ID,

editors. Hymenoptera and Biodiversity. Oxon: CAB International; 1993. pp. 53-82

402. Forister ML, Shapiro AM. Climatic trends and advancing spring flight of butterflies in lowland California. Global Change Biology. 2003;9:1130-1135

403. Davis AR. Influence of elevated CO_2 and ultraviolet-B radiation levels on floral nectar production: A nectary-morphological perspective. Plant Systematics and Evolution. 2003;238:169-181

404. Collins SA, Robinson GE, Conner JK. Foraging behavior of honey bees (Hymenoptera: Apidae) on Brassica nigraand B. rapagrown under simulated ambient and enhanced UV-B radiation. Annals of the Entomological Society of America. 1997;90(1):102-106

405. Sampson BJ, Cane JH. Impact of enhanced ultraviolet-B radiation on flower, pollen, and nectar production. American Journal of Botany. 1999;86:108-114

406. Crozier L. Winter warming facilitates range expansion: Cold tolerance of the butterfly Atalopedescampetris. Oecologia. 2003;135:648-656

407. Walther GR. Plants in a warmer world. Perspectives in Plant Ecology, Evolution and Systematics. 2004;6:169-185

408. Wilson R, Gutiérrez D, Gutiérrez J, Martínez D, Agudo R, Monserrat VJ. Changes to the elevational limits and extent of species ranges associated with climate change. Ecology Letters. 2005;8(11):1138

409. Bhagwat S. Ecosystem services and sacred natural sites: Reconciling material and non-material values in nature conservation. Environmental Values. 2009;18:417-427

410. Blicharska M, Mikusiński G, Godbole A, andSarnaik J. Safeguarding biodiversity and ecosystem services of sacred groves—Experiences from

northern Western Ghats. International Journal of Biodiversity Science, Ecosystem Services & Management. 2013;9:339-346

411. Bodin O, Tengo M, Norman M, Lundberg J, Elmqvis T. The value of small size: Loss of forest patches and ecological thresholds in Southern Madgascar. Ecological Applications. 2006;16:440-451

412. ICCA. Bio-cultural Diversity Conserved by Indigenous Peoples & Local Communities—Examples & Analysis. 2012. Available from: http://www.cenesta.org/wp-content/uploads/2015/10/publication-ceesp-briefing-note-10-companion-en.pdf

413. Batary P et al. Effect of conservation management on bees and insect-pollinated grassland plant communities in three European countries. Agriculture, Ecosystems and Environment. 2010;136:35-39

414. Morandin LA, Winston ML. Pollinators provide economic incentive to preserve natural land in agroecosystems. Agriculture Ecosystems and Environment. 2006;116:289-292

415. Cavigli I, Daughenbaugh KF, Martin M, Lerch M, Banner K, Garcia E, et al. Pathogen prevalence and abundance in honey bee colonies involved in almond pollination. Apidologie. 2016;47:251-266

416. Moritz RF, Härtel S, Neumann P. Global invasions of the western honeybee (Apismellifera) and the consequences for biodiversity. Ecoscience. 2005;12(3):289-301

417. Ritter W. Good beekeeping practice—Knowledge in a Nutshell. Bees for Development Journal. 2013;107:3-5

418. Roulston TH, Cane JH, Buchmann SL. What governs the protein content of pollen grains: pollinator preferences, pollen-pistil interactions, or phylogeny? Ecological Monographs. 2000;70:617-643

419. Shepherd MD, Buchmann SL, Vaughan M, Black SH. Pollinator Conservation Handbook: A Guide to Understanding, Protecting, and Providing Habitat for Native Pollinator Insects. The Xerces Society in Association with The Bee Works. Portland: The Xerces Society; 2003. 145 pp

420. Michener CD. Bees of the World. Baltimore and London: The Johns Hopkins University Press; 2000. 913 pp

421. Burkle L, Marlin J, Knight T. Plant-pollinator interactions over 120 years: Loss of species, co-occurrence, and function. Science. 2013;80:1611-1615

422. Gonzalez-Varo JP, Biesmeijer JC, Bommarco R, Potts SG, Schweiger O, Smith HG, et al. Combined effects of global change pressures on animal-mediated pollination. Trends in Ecology and Evolution. 2013;28:524-530

423. Fox R. The decline of moths in Great Britain: A review of possible causes. Insect Conservation and Diversity. 2012;6:5-19

424. Maes D, Van Dyck H. Butterfly diversity loss in Flanders (north Belgium): Europe‘s worst case scenario? Biological Conservation. 2001;99:263-276

425. Biesmeijer JC et al. Parallel declines in pollinators and insect pollinated plants in Britain and the Netherlands. Science. 2006;313:351-354

426. Kosior A et al. The decline of the bumble bees and cuckoo bees (Hymenoptera: Apidae: Bombini) of western and central Europe. Oryx. 2007;41:79-88

427. van Engelsdorp D et al. A survey of honey bee colony losses in the U.S., fall 2007 to spring 2008. PLoS ONE. 2008;3:e4071

428. Kevan PG. Pollination and environment conservation. Environmental Conservation. 1975;2:222-227

429. Keven PG. Pollination and flower visiting insects and the management of beneficial and harmful insects and plants. In: Hussein MY, Ibrahim AG, editors. Biological Control in the Tropics. Serdang, Selangor, Malaysia: University Pertanian Malaysia; 1986. pp. 439-452

430. Keven PG, editor. Alternative pollinators for Onario‘s crops. Proceedings of the Entomological Society of Ontario. 1987;118:109-170

431. Boyle-Makowski RMD. The importance of native pollinators in cultivated orchards: Their abundance and activities in relation to weather conditions. Proceedings of the Entomological Society of Ontario. 1987;118:125-141

432. Aguilar R, Ashworth L, Galetto L, Aizen MA. Plant reproductive susceptibility to habitat fragmentation: Review and synthesis through a meta-analysis. Ecology Letters. 2006;9:968-980

433. Montero-Castano A, Vila M. Impact of landscape alteration and invasions on pollinators: A meta-analysis. Journal of Ecology. 2012;100:884-893

434. Morales CL, Traveset A. A meta-analysis of impacts of alien vs. native plants on pollinator visitation and reproductive success of co-flowering native plants. Ecology Letters. 2009;12:716-728

435. Edge DA, Mecenero S. Butterfly conservation in Southern Africa. Journal of Insect Conservation. 2015;19:325-339. DOI: 10.1007/s10841-015-9758-5

436. Mecenero S, Ball JB, Edge DA, et al. Conservation Assessment of Butterflies of South Africa. Lesotho and Swaziland: Red List and Atlas; 2013

437. Nieto A, Roberts SPM, Kemp J, Rasmont P, Kuhlmann M, Biesmeijer JC, et al. European Red List of Bees. Luxembourg: Publication Office of the European Union; 2014

438. Anderson DL, Trueman JWH. Varroajacobsoni (Acari: Varroidae) is more than one species. Experimental and Applied Acarology. 2000;24:165-189
439. Sammataro D, Gerson U, Needham G. Parasitic mites of honey bees: Life history, implications, and impact. Annual Review of Entomology. 2000;45:519-548
440. McMahon DP, Fürst MA, Caspar J, Theodorou P, Brown MJF, Paxton RJ. A sting in the spit: Widespread cross-infection of multiple RNA viruses across wild and managed bees. The Journal of Animal Ecology. 2015;84:615-624

About Authors

Dr. Fadhil J. Kadhim

Horticulture and landscaping, College of Agriculture, University of Al-Qasim Green, Iraq

Prof. Dr. Mohamed Abdel-Raheem Ali Abdel-Raheem

Professor of Biological Control

Pests & Plant Protection Department

Agricultural and Biological Research Institute

National Research Centre.

Prof. Dr. Mohamed Abdel-Raheem Ali Abdel-Raheem, Prof. of Entomology (Biological Control), Pests & Plant Protection Department, Agricultural and Biological Research Institute, National Research Centre, Cairo, Egypt. Published publications (320) papers (13), Books and Book chapters (115), (Scopus) h-index (11), Citations (323), (Google Scholar) h-index (19) , Citations (1112), (Web of Science) Publications (28), h-index (10), Citations (285), Research Gate, h-index (14), Citations (679), Reviewers in International Journals (155), Reviewed articles in Biological Control (980), Editor in Chief in Journal (7), Assistant Editor (1), Associated Editor in Chief in Journal (7), Editorial board in Journals (42), Advisory board (1), conference Attended (27), workshop Attended (303), Symposium Attended (179), Forum Attended (4), Prize (4), Attended Others (30), Projects As PI and member (20), Training courses For Agric. Engineering (61), Training courses For Students of University (8), attending Trainees courses (19), TV, Radio & Newspaper (35), Supervisor on Ms.C. thesis (2), Supervisor on Ph.D. thesis (1), Committee Ph.D. thesis in Egypt (2), and member in scientific Foundation (14), Giving International lectures (4), Promotion Committee Prof. international (1), Committee Ph.D. thesis international (1).

National Research Centre. 33rd ElBohouth St., Dokki, Cairo, Egypt.

Email Address: abdelraheem_nrc@hotmail.com, ma.abdel-raheem@nrc.sci.eg, abdelraheem_nrc@yahoo.com, profmohamedabdelraheem@gmail.com, abdelraheem196942@gmail.com, mabdelraheem69@gmail.com, mabdelraheem69@yahoo.com.

Mobile Phone: (+2) 01155527583 – (+2) 01009580797

Scopus Author ID : 55220661700

https://www.scopus.com/authid/detail.uri?authorId=55220661700

http://scholar.google.com/citations?hl=en&user=gBtssEgAAAAJ

https://orcid.org/my-orcid?orcid=0000-0001-9240-064X
https://www.webofscience.com/wos/author/searchAbdel-Raheem, Mohamed Abdel-Raheem Ali - Web of Science Core Collection
https://www.researchgate.net/profile/Mohamed_Abdel-Raheem4/publications?sorting=recentlyAdded
http://livedna.org/20.16018
http://www.resarcherid.com/rid/R-6264-2017
https://www.webofscience.com/wos/author/record/R-6264-2017
https://www.facebook.com/share/p/1Piu6xyFmN/

- TV, Radio & Newspaper links

1. https://www.youtube.com/attribution_link?a=NsH2xGv9mY8&u=%2Fwatch%3Fv%3Dsb46rBG8-Mc%26feature%3Dshare
2. https://youtu.be/Ub8djW0tggo
3. https://m.facebook.com/story.php?story_fbid=492907069288338&id=100076011247165
4. https://youtu.be/dOujUAZmTdw
5. https://m.facebook.com/story.php?story_fbid=492907069288338&id=100076011247165
6. https://fb.watch/gPGrDuzZFX/
7. https://www.youtube.com/watch?si=4UfyQaeHVLPiVR5J&v=nXV4NT-no7o&feature=youtu.be
8. https://www.youtube.com/watch?v=sRH44W6TbcQ
9. https://youtu.be/uq9Qj3AEBHk?si=ymaX06huG2AXJESu
10. https://www.facebook.com/share/v/Qg5s6HW3hD8PLBzp/?mibextid=xfxF2i
11. https://el-yom.com/news/archives/1092767

12. https://el-yomnews.com/?p=1156515
13. https://el-yomnews.com/%D8%A7%D9%84%D8%AF%D9%83%D8%AA%D9%88%D8%B1-%D8%B9%D8%A8%D8%AF%D8%A7%D9%84%D8%B1%D8%AD%D9%8A%D9%85-%D9%85%D8%AD%D9%85%D8%AF-%D8%B9%D9%84%D9%8A-%D9%8A%D8%AA%D8%AD%D8%AF%D8%AB-%D9%84-%D9%84%D9%84/
14. https://el-yomnews.com/?p=1179515
15. https://www.facebook.com/share/v/4MM85wKRPxsYLqTr/
16. http://https//el-yomnews.com/%D8%A7%D9%84%D8%AD%D8%B4%D8%B1%D8%A7%D8%AA-%D9%81%D9%8A-%D8%AE%D8%AF%D9%85%D8%A9-%D8%A7%D9%84%D8%B9%D8%AF%D8%A7%D9%84%D8%A9-%D9%85%D8%AD%D8%A7%D8%B6%D8%B1%D8%A9-%D8%AA%D9%83%D8%B4%D9%81-%D8%AF%D9%88/
17. https://youtube.com/watch?v=F21ziQxdeco&si=LvSUixmnSg4JM1zK
18. https://youtu.be/F21ziQxdeco?si=jyAwbU8dVwkYzRH4

19- https://el-yomnews.com/ل/لماكاف حةالمجي يهتلفطرياتحلولهبتك
20- https://el-yomnews.com/archives/1296404
21- https://www.facebook.com/share/p/1AC25ue5XG/
22- https://youtu.be/7VItE5kON2g?si=9rMWMUSGk8Csgjfn
23- https://www.youtube.com/watch?v=G9nXA7Q9C60
24- https://www.youtube.com/watch?v=7VItE5kON2g
25- https://el-yomnews.com/archives/1347977

TV and Radio

1- Filming in the program (Meeting the Scientists) - Al-Manara Channel, entitled Biological Control of Insect Pests, on 7/2/2013.
2- Filming in the program (Friends of Al-Manara) - Al-Manara Channel, February 2014.
3- A recording of an interview with the program (From the Secrets of the Universe and Life) Radio Sawt Al-Arab, Cairo on 4/26/2015.
4- Filming in the program (Idea for a bobbin) - Nile Cultural Channel under the title ~~-U~~sing entomopathogenic fungi and plant extracts in combating insect pests that infect sugar crops in Egypt" on 11/1/2015.
5- Filming on (This Morning) program - Nile News Channel under the title: Submitting the specifications for the reclamation of one and a half million feddans as a first stage of the reclamation of 4 million feddans on 11/15/2015. https://www.youtube.com/attribution_link?a=NsH2xGv9mY8&u=%2Fwatch%3Fv%3Dsb46rBG8-Mc%26feature%3Dshare
6- Registration in the program (Our Life Between Our Hands) for North Upper Egypt Radio and Radio, presented by Dr. Essam Fathy, 3 episodes, on 11-29-2021, and it will be broadcast on Saturday and Wednesday at 3:30 pm in January 2022
7- Registration in the program (Our Life Between Our Hands) for North Upper Egypt Radio and Radio, presented by Dr. Essam Fathy, 3 episodes, on 11-29-2021, and it will be broadcast on Saturday and Wednesday at 3:30 pm in January 2022

8- Registration in the program (Our Life Between Our Hands) for North Upper Egypt Radio and Radio, presented by Dr. Essam Fathy, 3 episodes, on 11-29-2021, and it will be broadcast on Saturday and Wednesday at 3:30 pm in January 2022

9- Registration in ―Our Green Land" program for North Upper Egypt Radio and Radio, presented by Dr. Essam Fathy, 3 episodes, on 11-2021-2021, to be broadcast in January 2022

10- Registration in ―Our Green Land" program for North Upper Egypt Radio and Radio, presented by Dr. Essam Fathy, 3 episodes, on 11-2021-2021, to be broadcast in January 2022

11- Registration in ―Our Green Land" program for North Upper Egypt Radio and Radio, presented by Dr. Essam Fathy, 3 episodes, on 11-2021-2021, to be broadcast in January 2022

12- Filming in the program (Egypt is Beautiful) - the second Egyptian Nile Channel, under the title: Climate Changes on November 6, 2022. https://youtu.be/dOujUAZmTdw

13- Registration in ―Our Green Land" program for North Upper Egypt Radio and Radio, presented by Dr. Essam Fathy, 3 episodes, on 11-2021-2021, to be broadcast in January 2022

14- Filming in the program (Egypt and Climate Summit) - Delta Satellite Channel under the title: Climate Cop 27 on 13/11/2022 https://youtu.be/Ub8djW0tggo

15- Filming in (Egypt) - Sohag, Coral Times Hotel, under the title: Climate Change, Cop 27, dated 11/15/2022, the journalist Shorouk Abdullah https://m.facebook.com/story.php?story_fbid=492907069288338&id=100076011247165

16- Filming in (Feed and Benefit) - Coral Hotel in Sohag under the title: Climate Cop 27 on 11/15/2022, journalist Ahmed Tarboush https://fb.watch/gPGrDuzZFX/

17- An intervention in the plant clinic program on insect-pathogenic fungi in Egypt's agricultural farms on Thursday 29/2/2024.
https://www.youtube.com/watch?si=4UfyQaeHVLPiVR5J&v=nXV4NT-no7o&feature=youtu.be

18- Filming in the plant clinic program on the Misr Agricultural Channel about insect-pathogenic fungi on Sunday 3-3-2024
https://www.youtube.com/watch?v=sRH44W6TbcQ

19- Filming in the plant clinic program at the Misr Agricultural Channel, early prediction of the red palm weevil and modern methods of combating it, on Thursday 25/4/2024.
https://youtu.be/uq9Qj3AEBHk?si=ymaX06huG2AXJESu

20- Filming in the Green Planet program on the Egyptian Channel 2 (Biological Control) on Sunday 7/6/2024
https://www.facebook.com/watch/?v=981454710052376

21- A press interview in Al-Youm newspaper with journalist Mr. Imad Naseer on 7-30-2024. https://el-yom.com/news/archives/1092767
https://el-yomnews.com/

22- https://el-yomnews.com/?p=1156515

23- https://el-yomnews.com/%D8%A7%D9%84%D8%AF%D9%83%D8%AA%D9%88%D8%B1-%D8%B9%D8%A8%D8%AF%D8%A7%D9%84%D8%B1%D8%AD%D9%8A%D9%85-%D9%85%D8%AD%D9%85%D8%AF-

%D8%B9%D9%84%D9%8A-
%D9%8A%D8%AA%D8%AD%D8%AF%D8%AB-%D9%84-
%D9%84%D9%84/

24- https://el-yomnews.com/?p=1179515

25- https://www.facebook.com/share/v/4MM85wKRPxsYLqTr/

26- Insects in the Service of Justice: A Lecture Explaining the Role of Forensic Entomology in Investigating Crimes
https://el-yomnews.com/archives/1243894

27- Human health is at risk. An investigation reveals the secrets of pesticide-contaminated agriculture.
https://el-yomnews.com/archives/1296404

28- https://www.facebook.com/share/p/1AC25ue5XG/
The Green Planet Program (Beneficial Insects and Their Role in the Environment) on 5/23/2025 on the Egyptian Channel 2

29- https://youtu.be/7VItE5kON2g?si=9rMWMUSGk8Csgjfn

30- https://www.youtube.com/watch?v=G9nXA7Q9C60

31- https://www.youtube.com/watch?v=7VItE5kON2g

32- https://el-yomnews.com/archives/1347977

Assist. Teacher. Bayader Abd al-Jabbar Ali
Horticulture and landscaping , College of Agriculture, University of Al-Qasim Green, Iraq